D1051948

STEPHEN JAY GOULD

ROCKS OF AGES

The frontispiece shows the title-page illustration (modified only by a different title and author!) of the greatest geological treatise ever written by a scientist who also held holy orders—the Mundus subterraneus *(Underground World) by the great Jesuit scholar Athanasius Kircher, published in 1664. I regard the figure as a beautiful illustration of science and religion working together in their different ways. God holds the earth in space, but twelve winds in human form control both motion and climate, while the banner cites a famous line from Virgil's* Aeneid, *ending* mens agitat molem, *usually slightly mistranslated as "mind moves mountains"* (moles, accusative molem, *refers to any massive structure).*

ROCKS OF AGES

Science and Religion
in the Fullness of Life

STEPHEN JAY GOULD

BALLANTINE BOOKS · NEW YORK

A Ballantine Book
Published by The Ballantine Publishing Group

Copyright © 1999 by Stephen Jay Gould
Grateful acknowledgment is made to the following for permission to reprint
previously published material:

Henry Holt and Company, Inc., and Jonathan Cape Ltd.: "Design" from *The Poetry of
Robert Frost* edited by Edward Connery Lathem. Copyright © 1936 by Robert
Frost. Copyright © 1964 by Lesley Frost Ballantine. Copyright © 1969 by Henry
Holt & Company. Reprinted by permission of Henry Holt and Company, Inc.,
and Jonathan Cape Ltd., on behalf of the Estate of Robert Frost.

W.W. Norton & Company, Inc., and Random House UK Ltd.: Excerpt from *Bully for
Brontosaurus: Reflections in Natural History* by Stephen Jay Gould. Copyright © 1991
by Stephen Jay Gould. Reprinted by permission of W.W. Norton &
Company, Inc., and Hutchinson, Random House UK Ltd.

www.ballantinebooks.com

Cover painting: Stanze di Rafaello, *Adam and Eve*, c. 1509–1510.
Vatican Collection. Scala /Art Resource, N.Y.

Library of Congress Catalog Card Number: 2001119181

ISBN 0-345-45040-X

Manufactured in the United States of America

First Hardcover Edition: March 1999

First Trade Paper Edition: March 2002

10   9   8   7   6   5   4

For Jesse and Ethan,

who will have to hold on beyond their father's watch, and who will surely improve a world with a future so honestly described by John Playfair, a great scientist and writer, who closed his Outlines of Natural Philosophy *(1814) by stating (in the old subjunctive mood, where his "were" equals our "would be"):*

*"It were unwise to be sanguine,
and unphilosophical to despair."*

Contents

1

THE PROBLEM
STATED

Preamble

I WRITE THIS LITTLE BOOK to present a blessedly simple and entirely conventional resolution to an issue so laden with emotion and the burden of history that a clear path usually becomes overgrown by a tangle of contention and confusion. I speak of the supposed conflict between science and religion, a debate that exists only in people's minds and social practices, not in the logic or proper utility of these entirely different, and equally vital, subjects. I present nothing original in stating the basic thesis (while perhaps claiming some inventiveness in choice of illustrations); for my argument follows a strong consensus accepted for decades by leading scientific and religious thinkers alike.

Our preferences for synthesis and unification often prevent us from recognizing that many crucial problems in our complex lives find better resolution under the

opposite strategy of principled and respectful separation. People of goodwill wish to see science and religion at peace, working together to enrich our practical
and ethical lives. From this worthy premise, people
often draw the wrong inference that joint action implies
common methodology and subject matter—in other
words, that some grand intellectual structure will bring
science and religion into unity, either by infusing nature
with a knowable factuality of godliness, or by tooling up
the logic of religion to an invincibility that will finally
make atheism impossible. But just as human bodies require both food and sleep for sustenance, the proper
care of any whole must call upon disparate contributions from independent parts. We must live the fullness
of a complete life in many mansions of a neighborhood
that would delight any modern advocate of diversity.

I do not see how science and religion could be unified, or even synthesized, under any common scheme
of explanation or analysis; but I also do not understand
why the two enterprises should experience any conflict.
Science tries to document the factual character of the
natural world, and to develop theories that coordinate
and explain these facts. Religion, on the other hand,
operates in the equally important, but utterly different,
realm of human purposes, meanings, and values—subjects
that the factual domain of science might illuminate, but
can never resolve. Similarly, while scientists must oper-

ate with ethical principles, some specific to their practice, the validity of these principles can never be inferred from the factual discoveries of science.

I propose that we encapsulate this central principle of respectful noninterference—accompanied by intense dialogue between the two distinct subjects, each covering a central facet of human existence—by enunciating the Principle of NOMA, or Non-Overlapping Magisteria. I trust that my Catholic colleagues will not begrudge this appropriation of a common term from their discourse—for a magisterium (from the Latin *magister*, or teacher) represents a domain of authority in teaching.

Magisterium is, admittedly, a four-bit word, but I find the term so beautifully appropriate for the central concept of this book that I venture to impose this novelty upon the vocabulary of many readers. This request for your indulgence and effort also includes a proviso: Please do not mistake this word for several near homonyms of very different meaning—*majesty*, *majestic*, etc. (a common confusion because Catholic life also features activity in this different domain). These other words derive from the different root (and route) of *majestas*, or majesty (ultimately from *magnus*, or great), and do imply domination and unquestioning obedience. A magisterium, on the other hand, is a domain where one form of teaching holds the appropriate tools for meaningful discourse and resolution. In other words, we debate and hold dialogue under a magisterium; we fall into silent awe or imposed obedience before a majesty.

To summarize, with a tad of repetition, the net, or magisterium, of science covers the empirical realm: what is the universe made of (fact) and why does it work this way (theory). The magisterium of religion extends over questions of ultimate meaning and moral value. These two magisteria do not overlap, nor do they encompass all inquiry (consider, for example, the magisterium of art and the meaning of beauty). To cite the old clichés, science gets the age of rocks, and religion the rock of ages; science studies how the heavens go, religion how to go to heaven.

I will examine this NOMA principle as a solution to the false conflict between science and religion in four chapters: the first, an introduction based on two stories and contrasts; the second, a characterization and illustration of NOMA as developed and supported by both institutions of science and religion; the third, an outline of historical reasons for the existence of conflict, where none should exist; and the fourth, a summary of psychological reasons for the same false conflict, with a closing suggestion for the path of best interaction.

I deplore the current penchant for literary confession, spawned by our culture's conflation of two radically different concepts: celebrity and stature. Nonetheless, I accept that intellectual subjects of such personal salience impose some duty for authorial revelation—while the essay, as a literary genre, has been defined as discussion of general ideas in personal contexts ever since Montaigne coined the name in the sixteenth century. Let me, then, briefly state a perspective born of my own accidental ontogeny.

I grew up in an environment that seemed entirely conventional and uninteresting to me—in a New York Jewish family following the standard pattern of generational rise: immigrant grandparents who started in the sweatshops, parents who reached the lower ranks of the middle classes but had no advanced schooling, and my third generation, headed for a college education and a professional life to fulfill the postponed destiny. (I remember my incredulity when the spouse of an English colleague of "good breeding" found this background both exotic and fascinating. I also remember two incidents that emphasize the extreme parochiality of my apparent sophistication as a child on the streets of New York: First, when my father told me that Protestantism was the most common religion in America, and I didn't believe him because just about everyone in my neighborhood was either Catholic or Jewish—the composition of New York's rising Irish, Italian, and Eastern European working classes, the only world I knew. Second, when my one Protestant friend from Kansas City introduced me to his grandparents, and I didn't believe him—because they spoke unaccented English, and my concept of "grandparent" had never extended beyond European immigrants.) I had dreamed of becoming a scientist in general, and a paleontologist in particular, ever since the *Tyrannosaurus* skeleton awed and scared me at New York's Museum of Natural History when I was five years old. I had the great good fortune to achieve these goals and to love the work with fully sustained joy to this day, and without a moment of doubt or any extended boredom.

I shared the enormous benefit of a respect for learning that pervades Jewish culture, even at the poorest economic levels. But I had no formal religious education—I did not even have a bar mitzvah—because my parents had rebelled against a previously unquestioned family background. (In my current judgment, they rebelled too far, but opinions on such questions tend to swing on a pendulum from one generation to the next, perhaps eventually coming to rest at a wise center.) But my parents retained pride in Jewish history and heritage, while abandoning all theology and religious belief. (The Holocaust claimed most of both sides of my family—nothing directly personal, for I knew none of these relatives—so denial and forgetfulness could not have been an option for my parents.)

I am not a believer. I am an agnostic in the wise sense of T. H. Huxley, who coined the word in identifying such open-minded skepticism as the only rational position because, truly, one cannot know. Nonetheless, in my own departure from parental views (and free, in my own upbringing, from the sources of their rebellion), I have great respect for religion. The subject has always fascinated me, beyond almost all others (with a few exceptions, like evolution, paleontology, and baseball). Much of this fascination lies in the stunning historical paradox that organized religion has fostered, throughout Western history, both the most unspeakable horrors and the most heartrending examples of human goodness in the face of personal danger. (The evil, I believe, lies in the frequent confluence of religion with secular power. Christianity has sponsored its share of horrors, from inquisitions to liquidations—but only because this institution held great secular power during so much of Western history. When my folks held such sway, more briefly and in Old Testament time, we committed similar atrocities with the same rationales.)

I believe, with all my heart, in a respectful, even loving, concordat between the magisteria of science and religion—the NOMA concept. NOMA represents a principled position on moral and intellectual grounds, not a merely diplomatic solution. NOMA also cuts both ways. If religion can no longer dictate the nature of factual conclusions residing properly within the magisterium of science, then scientists cannot claim higher insight into moral truth from any superior knowledge of the world's empirical constitution. This mutual humility leads to important practical consequences in a world of such diverse passions. We would do well to embrace the principle and enjoy the consequences.

A Tale of Two Thomases

THE DISCIPLE THOMAS MAKES THREE prominent appearances in the Gospel of John, each to embody
an important moral or theological principle. Nonetheless, these three episodes cohere in an interesting
way that can help us to understand the different powers
and procedures of science and religion. We first meet
Thomas in chapter 11. Lazarus has died, and Jesus
wishes to return to Judaea in order to restore his dear
friend to life. But the disciples hesitate, reminding Jesus
of the violent hostility that had led to a stoning on his
last visit. Jesus, in his customary manner, tells an ambiguous little parable, ending with the firm conclusion
that he will and must go to Lazarus—and Thomas steps
forth to break the deadlock and restore courage to
the disciples: "Then said Thomas . . . unto his fellow-
disciples, Let us also go, that we may die with him."

In the second incident (chapter 14), Jesus, at the Last Supper, states that he will be betrayed, and must endure bodily death as a result. But he will go to a better place and will prepare the way for his disciples: "In my Father's house are many mansions . . . I go to prepare a place for you." Thomas, now confused, asks Jesus: "Lord, we know not whither thou goest; and how can we know the way?" Jesus responds in one of the most familiar Bible passages: "I am the way, and the truth, and the life: no one cometh unto the Father, but by me."

According to legend, Thomas led a brave life after the death of Jesus, extending the gospel all the way to India. The first two biblical incidents, cited above, also display his admirable qualities of bravery and faithful inquiry. Yet we know him best by the third tale, and by an appended epithet of criticism—for he thus became the Doubting Thomas of our languages and traditions. In chapter 20, the resurrected Jesus appears first to Mary Magdalene, and then to all the disciples but the absent Thomas. The famous tale unfolds:

But Thomas was not with them when Jesus came. The other disciples therefore said unto him, We have seen the Lord. But he said unto them, Except I shall see in his hands the print of the nails,

and put my finger into the print of the nails, and
thrust my hand into his side, I will not believe.

Jesus returns a week later to complete the moral tale
of a brave and inquisitive man, led astray by doubt, but
chastened and forgiven with a gentle but firm lesson for
us all:

> Then came Jesus, the doors being shut, and
> stood in the midst and said, Peace be unto you.
> Then saith he to Thomas, Reach hither thy fin-
> ger, and behold my hands; and reach hither thy
> hand, and thrust it into my side: and be not
> faithless, but believing. And Thomas answered
> and said unto him, My Lord and my God.

(This last passage assumes great importance in traditional exegesis as representing the first time that a disciple identifies Jesus as God. Trinitarians point to
Thomas's utterance as proof for the threefold nature of
God as Father, Son, and Holy Ghost at the same time.
Unitarians must work their way around the literal
meaning, arguing, for example, that Thomas had
merely uttered an oath of astonishment, not an identifi-
cation.) In any case, Jesus' gentle rebuke conveys the
moral punch line, and captures the fundamental difference between faith and science:

Jesus saith unto him, Thomas, because thou hast seen me, thou hast believed: blessed are they that have not seen, and yet have believed.

Thomas, in other words, passes his test because he accepted the evidence of his observations and then repented his previous skepticism. But his doubt signifies weakness, for he should have known through faith and belief. The Gospel text emphasizes Thomas's failings through his exaggerated need to see *both* sets of stigmata (hands and side), and use *two* senses (sight and touch) to assuage his doubts.

Mark Tansey, a contemporary artist who loves to represent the great moral and philosophical lessons of Western history with modern metaphors painted in hyperrealistic style, beautifully epitomized the overly wrought character of Thomas's doubt. In 1986 he depicted a man who won't accept continental drift in general, or even the reality of earthquakes in particular. An earthquake has fractured both a California road and the adjoining cliff, but the man still doubts. So he instructs his wife, at the wheel, to straddle the fault line with their car, while he gets out and thrusts his hand into the analogy of Christ's pierced side—the crack in the road. Tansey titles this work *Doubting Thomas*.

I accept the moral of this tale for important principles under the magisterium of ethics and values. If you

Doubting Thomas by Mark Tansey (courtesy of the Curt Marcus Gallery)

need to go through the basic argument, and to test the consequences, each time anger tempts you to murder, then your fealty to the Sixth Commandment is a fragile thing indeed. The steadfast, in such cases, are more blessed (and more to be trusted) than those who cavil and demand rationales each time. Blessed are they that have no such need, yet know the way of justice and decency. In this sense, Thomas deserved his chastening—while Jesus, through the firm gentleness of his rebuke, becomes a great teacher.

But I cannot think of a statement more foreign to the norms of science—indeed more unethical under this magisterium—than Jesus' celebrated chastisement of Thomas: "blessed are they that have *not* seen, and yet have believed." A skeptical attitude toward appeals based only on authority, combined with a demand for direct evidence (especially to support unusual claims), represents the first commandment of proper scientific procedure.

Poor Doubting Thomas. At his crucial and eponymous moment, he acted in the most admirable way for one style of inquiry—but in the wrong magisterium. He espoused the key principle of science while operating within the different magisterium of faith.

So if Thomas the Apostle defended the norms of science in the wrong magisterium of faith, let us consider another Thomas usually (but falsely) regarded as equally incongruous in the other direction—as a man of dogmatic religion who improperly invaded the magisterium of science. The Reverend Thomas Burnet (1635–1715), although unknown outside professional circles today, wrote one of the most influential books of the late seventeenth century—*Telluris theoria sacra*, or *The Sacred Theory of the Earth*, a work in four sections, with part one on the deluge of Noah, part two on the preceding paradise, part three on the forthcoming "burning of the world," and part four "concerning the new heavens and new earth," or paradise regained after the conflagration. This book not only became a "best-seller" in its own generation, but gained lasting fame as a primary inspiration (largely, but not entirely, in criticism) for two of the greatest and most comprehensive works of eighteenth-century intellectual history—the *Scienza nuova (New Science)* of Giambattista Vico in 1725, the foundation for historical studies of cultural anthropology, and the *Histoire naturelle* of Georges Buffon, the preeminent compendium of the natural world, begun in 1749.

But modern scientists dismiss Burnet as either a silly fool or an evil force who tried to reimpose the unquestionable dogmas of scriptural authority upon the new paths of honest science. The "standard" early history of geology, Archibald Geikie's *Founders of Geology* (1905 edition) featured Burnet's book among the "monstrous doctrines" that infected late-seventeenth-century science. One modern textbook describes Burnet's work as "a series of queer ideas about earth's development," while another dismisses the *Sacred Theory* as a "bizarre freak of pseudo-science."

Of course, Burnet did not operate as a modern scientist, but he faithfully followed the norms of his time for proper residence within the magisterium of scientific inquiry. Burnet did begin by assuming that the Bible told a truthful story about the history of the earth, but he did not insist on literal accuracy. In fact, he lost his prestigious position as private confessor to King William III for espousing an allegorical interpretation of creation as described in the Book of Genesis—for he argued that God's six "days" might represent periods of undetermined length, not literal intervals of twenty-four hours or physical episodes of one full rotation about an axis.

Burnet accepted the scriptural account as a rough description of actual events, but he insisted upon one principle above all: the history of the earth cannot be regarded as adequately explained or properly interpreted until all events can be rendered as necessary consequences of invariable natural laws, operating with the knowable regularity recently demonstrated for gravity and other key phenomena by his dear friend Isaac Newton. Ironically, the most bizarre features of Burnet's particular account arise from his insistence upon natural law as the source and explanation of all historical events in the earth's history—a difficult requirement given the peculiar and cataclysmic character of several biblical tales, including universal floods and fires.

Burnet begins, for example, by seeking a source for the water of Noah's flood. (He greatly underestimated the depth and extent of the earth's oceans, and therefore believed that present seas could not cover the mountains. "I can as soon believe," he wrote, "that a man could be drowned in his own spittle as that the world should be deluged by the water in it.") But Burnet then rejects, as outside his chosen magisterium of "natural" (i.e., scientific) explanation, the easiest and standard solution of his age: that God simply made the extra water by miraculous creation. For miracle, defined as divine suspension of natural law, must lie outside the compass of scientific explanation. Invoking the story of Alexander and the Gordian Knot, Burnet rejected this "easy way" as destructive of any scientific account. (According to legend, when Alexander the Great captured Gordium, the capital of Phrygia, he encountered a famous chariot, lashed to a pole with a knot of astonishing complexity. He who could untie the knot would conquer all Asia. So Alexander, using raw power to circumvent the rules of the game, took his sword and severed the knot clean through. Some call it boldness; I, and apparently Burnet as well, call it anti-intellectualism.) Burnet wrote:

> They say in short, that God Almighty created waters on purpose to make the deluge, and then annihilated them again when the deluge was to cease; and this, in a few words, is the whole

account of the business. This is to cut the knot when we cannot loose it.

Instead, Burnet devised a wonderfully wacky theory about a perfectly spherical original earth with a smooth and solid crust of land covering a layer of water below (the natural and eventual source of Noah's flood). This crust gradually dries and cracks; waters rise through the cracks and form clouds; the rains arrive and seal the cracks; the pressure of water rising from below finally bursts through the crust, causing the deluge and producing the earth's present rough topography. Wacky indeed, but fully rendered by natural law, and therefore testable and subject to disproof under the magisterium of science. Indeed, we have tested Burnet's ideas, found them both false and bizarre, and expunged his name from our pantheon of scientific heroes. But if he had simply advocated a divine creation of water, such a conventional and nonoperational account could never have inspired Buffon, Vico, and a host of other scholars.

Burnet followed the common view of a remarkable group of men, devout theists all, who set the foundations of modern science in late-seventeenth-century Britain—including Newton, Halley, Boyle, Hooke, Ray, and Burnet himself. Invoking a convenient trope of English vocabulary, these scientists argued that God would permit no contradiction between his *words* (as recorded in scripture) and his *works* (the natural world). This principle, in itself, provides no rationale for science, and could even contradict my central claim for science and religion as distinct magisteria—for if works (the natural world) must conform to words (the scriptural text), then doesn't science become conflated with, constrained by, and subservient to religion? Yes, under one possible interpretation, but not as these men defined the concept. (Always look to nuance and actual utility, not to a first impression about an ambiguous phrase.) God had indeed created nature at some inception beyond the grasp of science; but he also established invariant laws to run the universe without interference forever after. (Surely omnipotence must operate by such a principle of perfection, and not by frequent subsequent correction, i.e., by special miracle, to fix some unanticipated bungle or wrinkle—to make extra water, for example, when human sin required punishment.)

Thus, nature works by invariant laws subject to scientific explanation. The natural world cannot contradict scripture (for God, as author of both, cannot speak against himself). So—and now we come to the key point—if some contradiction seems to emerge between a well-validated scientific result and a conventional reading of scripture, then we had better reconsider our exegesis, for the natural world does not lie, but words can convey many meanings, some allegorical or metaphorical. (If science clearly indicates an ancient world, then the "days" of creation must represent periods longer than twenty-four hours.) In this crucial sense, the magisteria become separate, and science holds sway over the factual character of the natural world. A scientist may be pious and devout—as all these men were, with utmost sincerity—and still hold a conception of God (as an imperial clockwinder at time's beginning in this version of NOMA) that leaves science entirely free in its own proper magisterium.

I choose Thomas Burnet to illustrate this central principle for three reasons: (1) he was an ordained minister by primary profession (thereby illustrating NOMA if he truly kept these worlds distinct); (2) his theory has become an unfair source of ridicule under the fallacious notion that science must be at war with religion; and (3) he upheld the primacy of science in a particularly forceful way (and with even more clarity than his friend Isaac Newton, as we shall see on page 87). Recognizing the primacy of science in its proper magisterium, Burnet urges his readers not to assert a scriptural interpretation contrary to a scientific discovery, but to reexamine scripture instead—for science rules the magisterium of factual truth about nature:

'Tis a dangerous thing to engage the authority
of scripture in disputes about the natural world,

in opposition to reason; lest time, which brings all things to light, should discover that to be evidently false which we had made scripture assert.

In a lovely passage equating an independent magisterium for science with a maximally exalted concept of God, Burnet develops a striking metaphor for contrasting explanations of the earth's destruction in Noah's flood: do we not have greater admiration for a machine that performs all its appointed tasks (both regular and catastrophic) by natural laws operating on a set of initial parts, than for a device that putters along well enough in a basic mode, but requires a special visit from its inventor for anything more complex:

We think him a better artist that makes a clock that strikes regularly at every hour from the springs and wheels which he puts in the work, than he that so made his clock that he must put his finger to it every hour to make it strike: and if one should contrive a piece of clock-work so that it should beat all the hours, and make all its motions regularly for such a time, and that time being come, upon a signal given, or a spring touched, it should of its own accord fall all to pieces; would not this be looked upon as a piece of greater art, than if the workman came at that

time prefixed, and with a great hammer beat it
into pieces?

As a professional clergyman and a leading scientist,
Burnet practiced in both magisteria, and kept them separate. He allocated the entire natural world to science,
but he also knew that this style of inquiry could not
adjudicate issues beyond the power of factual information to illuminate, and in realms where questions of
natural law do not arise. Using an image from his own
century (we would define the boundaries differently today), Burnet grants the entire history of the earth to
science, but recognizes that any time before the creation
of matter, and any history after the Last Judgment, cannot be encompassed within the magisterium of natural
knowledge:

Whatsoever concerns this sublunary world in
the whole extent of its duration, from the
Chaos to the last period, this I believe Providence hath made us capable to understand . . .
On either hand is Eternity, before the World
and after, which is without [that is, outside of]
our reach: But that little spot of ground that lies
betwixt those two great oceans, this we are to
cultivate, this we are masters of, herein we are
to exercise our thoughts [and] to understand.

I may be reading too much into Burnet's words, but do I not detect a preference, or at least a great fondness, for the factuality of science when, in the chronological narrative of his *Sacred Theory of the Earth*, Burnet must bid adieu to reason as his guide, as he passes from the factually knowable history of an earth fully governed by natural law to a radically different future at the Last Judgment, when God will institute a new order, and can therefore only inform us (if at all) through the revelation of his words? Burnet speaks to the muse of science:

> Farewell then, dear friend, I must take another guide: and leave you here, as Moses upon Mount Pisgah, only to look into that land, which you cannot enter. I acknowledge the good service you have done, and what a faithful companion you have been, in a long journey: from the beginning of the world to this hour . . . We have travelled together through the dark regions of a first and second chaos: seen the world twice shipwrecked. Neither water nor fire could separate us. But now you must give place to other guides.

I told this tale of two Thomases to sharpen the distinctions between two entirely different but equally vital magisteria of our rich and complex lives—the two rocks of ages in my title. One must not assume that a book (the Bible in this case) or a day job (as a clergyman in this example) defines a magisterium. We must look instead to the subject, the logic, and the particular arguments. Our goal of mutual respect requires mutual understanding most of all. But I must complete this intuitive and particular case for NOMA by telling another story—with a similar message, but from the moral side this time—before presenting the more formal argument in chapter 2.

The Fate of Two Fathers

ICAN HARDLY THINK OF a more common, or sillier, fallacy of human thought and feeling than our propensity to construct "golden age" myths about a simpler past of rustic bliss. When I hear such reveries, I feel an almost irresistible urge to chime in with a grounding statement that should be chiseled into everyone's consciousness in uppercase as the Great Reminder. I don't like boom-box rappers, the U.S. tax code, or the corps of paparazzi any more than the next upright fellow; and I have often dreamed about making my fortune by marketing moral fiber in cereal boxes as a replacement for a seriously eroding natural product. But if anyone tells me that he would rather have lived a century ago, I will simply remind him of the one irrefutable trump card for choosing right now as the best world we have ever known: thanks to modern medicine, people of adequate means in the industrial world will probably enjoy a privilege never before vouchsafed to any human group. Our children will grow up; we will not lose half or more of our offspring in infancy or childhood. We will not have to sing Mahler's heartrending *Kindertotenlieder*, or "Songs on the Death of Children." We will not have to hire the local daguerreotypist to make the only image of our dead child. (Young children could rarely sit still long enough for the several minutes required by this early photographic technique. But the dead do not move, and many daguerreotypists specialized in this profitable, if ghoulish, work.)

Knowledge of the probabilities may have softened the blow, but an abstract understanding that half might die could hardly provide solace against the pain of losing a beloved individuality forever. Thus, our ancestors suffered—all of them, including kings and queens, industrial magnates and country squires, for wealth bought little respite when even the best doctors could do next to nothing.

The two greatest Victorian heroes in my profession of evolutionary biology, Charles Darwin and Thomas Henry Huxley, both blessed with more than adequate means and medical knowledge, lost their favorite children under the most painful of circumstances. Both men have served as the principal *bêtes noirs* of religious revivalists and fundamentalists ever since—Darwin simply for developing evolutionary theory, Huxley for his more active "parson baiting." (In a famous aphorism, Huxley said that he could never recall which side of the heart holds the mitral valve, named for a resemblance in shape to a bishop's hat—until he remembered "that a bishop's never in the right," and he then always knew that the mitral valve connects the left auricle and ventricle of the heart.) For both men, the deaths coincided with an intense dialogue that confronted their losses with traditional Christian sources of solace—and both men rejected the conventional comfort in a moving and principled manner.

One might therefore suppose that both men became embittered by a perceived hypocrisy (or at least a false hope) offered by a hidebound doctrine. Did those tragic and senseless deaths lead Darwin and Huxley to become the forthright foes of religion that our cardboard histories often portray, and that the model of intrinsic warfare between science and theology would anticipate? In fact, nothing so simple occurred—as both men showed only the dignity of their stature, and the subtlety of their intellect. Huxley and Darwin did indeed lose any vestige of a lingering personal belief in an intrinsically just world, governed by a loving anthropomorphic deity. But the pain of their personal losses only sharpened their understanding of the differences between science and religion, the respect due to both institutions when properly pursued in their respective magisteria, and the distinctions between questions that could be answered and those beyond our power to comprehend or even to formulate.

A familiar story proclaims that Darwin had planned a career as a "country parson" when he set sail on the *Beagle* voyage around the world, and eventually got sidetracked into another profession. But the common inference that Darwin's discovery of evolution led him both to apostasy and to a biological career cannot be sustained. In truth, Darwin had never been personally committed to theology as a calling. As a young man, his religious views remained decidedly lukewarm and passively conventional, simply because he had never given the matter any extensive thought. His designs on a parsonage arose more from an absence of alternative plans than from any active belief or desire. I strongly suspect that, as the Reverend Charles, he would have treated his day job in a hallowed and traditional manner among clerical naturalists—as a sinecure with adequate salary and minimal duties, leaving him adequate time and opportunity to follow his true bliss: collecting and publishing books on beetles, and other subjects in natural history.

Thus, as Darwin approached midlife in the tranquillity of substantial means, an excellent professional reputation, and a happy family in a country home, he had never struggled deeply with questions of personal religious belief, even though his evolutionary views had led him to question and abandon several traditional dogmas of his Anglican upbringing. But then, in a fateful interval between late 1850 and April 23, 1851, intellectual doubt and personal tragedy combined to change his world forever.

As Darwin finished several years of intense technical work on the taxonomy of barnacles, his precarious health also improved substantially, and he found himself with both time to read and tranquillity to contemplate. He decided, finally, to examine his own religious beliefs in a careful and systematic way. Darwin therefore turned to the work of a fascinating thinker, then all the rage, but unknown today, especially because his far more famous brother followed a different path and eclipsed him. The Newman brothers could not abide the inconsistencies that they detected in Anglican practice and belief. John Henry Newman created one of the biggest stirs in nineteenth-century British intellectual life by converting to Catholicism, and eventually becoming a cardinal. (Catholic students' organizations on American campuses are usually called Newman Societies in his honor.)

Francis William Newman, the cardinal's younger brother, graduated from Oxford with a higher degree and more potential promise as a future don, but left this pillar of the intellectual establishment for an eventual job as professor of Latin at the upstart and unorthodox University College, London, because he would not subscribe (as law and tradition then required for Oxbridge dons) to the Thirty-Nine Articles of Anglicanism. Newman then undertook a spiritual journey, through several popular books, to a position of intense religious belief, but based on rejection of dogmas and harsh traditional doctrines (particularly the idea of later reward or eternal punishment for earthly deeds)—all in favor of a system consistent with rational thought and the findings of modern science. With his usual intensity, Darwin studied all of Newman's major works between 1850 and 1851, reaching similar conclusions about the vacuousness (and often the cruelty as well) of traditional dogmas, but finding no solace in Newman's ideas about personal devotion, and therefore ending with skepticism toward all aspects of religious belief.

Darwin's scrutiny of Newman's works might not have affected his view of life so profoundly if his greatest personal tragedy had not unfolded at the same time. Darwin loved his eldest daughter, Annie, with a fierce tenderness inspired by a complex mixture of Annie's own sweet disposition and her resemblance to Charles's sister Susan, who had acted as a surrogate after the early death of Charles's mother, and who had so tenderly cared for Charles's father to the day of his death, just two years before. But Annie had always been a sickly child.

In March of 1851, Annie became so ill that Charles and his wife, Emma, decided to send the ten-year-old girl to Dr. Gully's clinic in Malvern, where Charles's own health had been so dramatically improved by the doctor's celebrated "water treatment." Annie would have her sister and a nurse for support and comfort. Charles accompanied the entourage to Malvern and stayed for several days. (Emma, in the late stages of pregnancy, followed the customs of her time, and remained "confined" at the Darwin country home.)

Annie prospered at first, but soon became violently ill. Charles hurried to her bedside, and spent several days in agonizing torture, as Annie rallied to hope, sank into despair, and eventually died on April 23. Charles wrote to his brother Erasmus: "God knows we can neither see on any side a gleam of comfort." A week later, he penned a private memoir on his sadness, and on Annie's lost beauty of body and soul: "Oh that she could now know how deeply, how tenderly we do still and shall ever love her dear joyous face. Blessings on her."

Annie's cruel death catalyzed all the doubts that Charles's reading of Newman, and his deeper scrutiny of religion, had engendered. He had permanently lost all personal belief in a caring God, and would never again seek solace in religion. He carefully avoided any direct statement in both his public and private writings, so we do not know his inner resolutions. I suspect that he accepted Huxley's dictum about agnosticism as the only intellectually valid position, while privately embracing a strong (and, as he well knew, quite unprovable) suspicion of atheism, galvanized by Annie's senseless death.

But if science and religion wage constant battle for the same turf, then Darwin should have become hostile and dismissive toward religion, and cynical about life in general. He should have wielded evolution as a bludgeon against false comfort and cruel deception in a world filled with the deaths of children and other heart-wrenching tragedies of no conceivable moral meaning. But Darwin took no such position. He grieved as deeply as any man ever has, and he worked his way through. He retained his zest for life and learning, and he rejoiced in the warmth and successes of his family. He lost personal comfort and belief in the conventional practice of religion, but he developed no desire to urge such a view upon others—for he understood the difference between factual questions with universal answers under the magisterium of science, and moral issues that each person must resolve for himself. He would fight fiercely for the truth of evolution's factuality, but the causes of life's history could not resolve the riddles of life's meaning. A knowledge of the medical causes of death could prevent future tragedies, but could never assuage the pain of immediate loss, or enlighten the general meaning of suffering.

We shall return to Darwin's remarkable letter to the Harvard botanist Asa Gray (who accepted natural selection and evolution, but urged Darwin to view such laws as instituted by God for a discernible purpose)—for I regard this document as the finest comment ever written on the proper relation of science and religion. But for now I cite his views of May 1860—nine years after Annie's death, and six months after publication of the *Origin of Species*—on why the factuality of evolution cannot resolve religious questions of ultimate meaning:

> With respect to the theological view of the question. This is always painful to me. I am bewildered. I had no intention to write atheistically. But I own that I cannot see as plainly as others do, and as I should wish to do, evidence of design and beneficence on all sides of us. There seems to me too much misery in the world . . . On the other hand, I cannot anyhow be contented to view this wonderful universe, and especially the nature of man, and to conclude that everything is the result of brute force. I am inclined to look at everything as resulting

from designed laws, with the details, whether good or bad, left to the working out of what we may call chance. Not that this notion at all satisfies me. I feel most deeply that the whole subject is too profound for the human intellect. A dog might as well speculate on the mind of Newton.

Thomas Henry Huxley, Darwin's brilliant and articulate younger colleague, and his "bulldog" in public support for evolution against all tides of social and religious orthodoxy, lost his favorite and firstborn son, three-year-old Noel, on September 15, 1860—just four months after Darwin's letter to Gray, and a year after Huxley himself had read the *Origin of Species* and exclaimed in awed astonishment, mixed with a twinge of envy, "how extremely stupid not to have thought of that!"

Darwin's Annie had been a sickly child, and her death fulfilled an acknowledged probability that Charles and Emma had hoped, above any other conceivable prayer, to forestall. But Huxley's rambunctious Noel had romped with his father before bedtime on Thursday, and then died on Saturday. "It was as if the boy had been inoculated with some septic poison," Huxley wrote. Among the many friends who tried to console him in this wrenchingly sudden and maximal grief, Huxley bared his soul in return only to the man he respected most highly and disagreed with most profoundly—the liberal clergyman Charles Kingsley, also an amateur naturalist of note, an evolutionist who saw no conflict between science and his ecclesiastical duties, and the popular author of *Westward Ho!* and *The Water Babies*.

Kingsley had reached out to his skeptical friend with a gentle suggestion that, in this hour of greatest conceivable need, Huxley might reexamine his doubts and find comfort in the Christian doctrine of eternal souls, and the attendant prospect that he would meet Noel again in a different life to come. In his letter, Kingsley acknowledges Huxley's suffering as "something horrible, intolerable, like being burnt alive." But we can find extended solace in earthly preparation for a heavenly meeting after bodily death. We must, Kingsley wrote, "make ourselves worthy for the reunion" during our sojourn on earth.

Huxley answered Kingsley in a letter of September 23, 1860, that should be required reading for all courses in English literature and philosophy. Great and passionate writing does not only appear in novels. As prose stylists, a few nineteenth-century scientists (Playfair, Lyell, and Huxley in particular) rank with the finest Victorian fiction writers. I wish I could quote this long letter in full, for I have never read a more moving or

incisive defense of personal intellectual honesty, whatever the allure of immediate and easy solace from com-
forts that one can neither truly believe nor justify by
cogent argument.

Huxley begins by thanking Kingsley for his proffered comfort, expressed with complete sincerity and
no sanctimoniousness. But Huxley explains, in a beautiful passage, that he cannot overthrow a personal philosophy, worked out over so many years and through so
much intellectual struggle, for the immediate solace of
a rejected central belief in the immortality of souls:

> My Dear Kingsley—I cannot sufficiently thank
> you, both on my wife's account and my own,
> for your long and frank letter, and for all the
> hearty sympathy which it exhibits . . . My con-
> victions, positive and negative, on all the matters
> of which you speak, are of long and slow
> growth and are firmly rooted. But the great
> blow which fell upon me seemed to stir them
> from their foundation, and had I lived a couple
> of centuries earlier I could have fancied a devil
> scoffing at me and them—and asking me what
> profit it was to have stripped myself of the
> hopes and consolations of the mass of mankind?
> To which my only reply was and is—Oh devil!
> truth is better than much profit. I have searched

over the grounds of my belief, and if wife and child and name and fame were all to be lost to me one after the other as the penalty, still I will not lie.

Huxley then epitomizes his arguments for skepticism about immortality: Why, first of all, should we grant eternity to complex humans and not to "lower" creatures who might benefit even more by such a blessing; and, second, why should we believe a doctrine mainly because we long so deeply for its validity?

Nor does the infinite difference between myself and the animals alter the case. I do not know whether the animals persist after they disappear or not. I do not even know whether the infinite difference between us and them may not be compensated by *their* persistence and *my* cessation after apparent death, just as the humble bulb of an annual lives, while the glorious flowers it has put forth die away.

Surely it must be plain that an ingenious man could speculate without an end on both sides, and find analogies for all his dreams. Nor does it help me to tell me that the aspirations of mankind—that my own highest aspirations even—lead me towards the doctrine of immortality. I doubt the fact, to begin with, but if it be so even, what is this but in grand words asking me to believe a thing because I like it.

Huxley then states his reasons for embracing science as his guide in factual questions. In the "standard quote" from this letter, the lines found in every *Bartlett's* ever published, Huxley writes:

My business is to teach my aspirations to conform themselves to fact, not to try and make facts harmonize with my aspirations. Science seems to me to teach in the highest and strongest manner the great truth which is embodied in the Christian conception of entire surrender to the will of God. Sit down before fact as a little child, be prepared to give up every preconceived notion, follow humbly wherever and to whatever abysses nature leads, or you shall learn nothing. I have only begun to learn content and peace of mind since I have resolved at all risks to do this.

These statements might be—and usually have been—taken as a manifesto for the standard model of warfare between science and religion, and as a classical defense for science, even in the hour of greatest spiritual need. But this wonderful letter, read *in extenso*, takes an opposite position, akin to Darwin's at the death of Annie. Huxley does reject the soul's immortality as a personal comfort in his grief—for all the reasons cited above. But he forcefully recognizes the major principle of NOMA in stating that such a religious idea cannot be subject to scientific proof: "I neither deny nor affirm the immortality of man. I see no reason for believing in it, but, on the other hand, I have no means of disproving it." Then, in terms strikingly similar to Darwin's metaphor about dogs and the mind of Newton (see page 36), Huxley locates this subject beyond the magisterium of science, and in the domain of personal decision, because we cannot even imagine a rational test: "in attempting even to think of these questions, the human intellect flounders at once out of its depth."

Then, in a concluding passage that still brings tears to my eyes, Huxley summarizes a personal case for NOMA by stating the three non-overlapping aspects of personal integrity—religion for morality, science for factuality, and love for sanctity—that have anchored his own life and given it meaning. He begins by quoting the philosophical work of his friend Thomas Carlyle (*Sartor Resartus*, or *The Tailor Reclothed*), and ends with the celebrated line of Martin Luther at the Diet of Worms, stating why he will not renounce his religious convictions: "God help me, I cannot do otherwise."

Has any "atheist" ever presented a better case for the role of true religion (as a ground for moral contemplation, rather than a set of dogmas accepted without questioning)?

> *Sartor Resartus* led me to know that a deep sense of religion was compatible with the entire absence of theology. Secondly, science and her methods gave me a resting-place independent of authority and tradition. Thirdly, love opened up to me a view of the sanctity of human nature, and impressed me with a deep sense of responsibility.
>
> If at this moment I am not a worn-out, debauched, useless carcass of a man, if it has been or will be my fate to advance the cause of science, if I feel that I have a shadow of a claim on the love of those about me, if in the supreme moment when I looked down into my boy's grave my sorrow was full of submission and without bitterness, it is because these agencies have worked upon me, and not because I have ever cared whether my poor personality shall remain distinct forever from the All from whence it came and whither it goes.
>
> And thus, my dear Kingsley, you will understand what my position is. I may be quite

wrong, and in that case I know I shall have to pay the penalty for being wrong. But I can only say with Luther, "Gott helfe mir, Ich kann nichts anders."

As a coda to this chapter, a symbolic story about Darwin's funeral, and Huxley's role in switching the intended place of burial, serves as a fitting symbol and illustration of NOMA, the potential harmony through difference of science and religion, both properly conceived and limited. Darwin wished to be buried in the local churchyard of his adopted village in Downe, where he had done the requisite good deeds for a man of wealth and social standing—including service as a magistrate, proper donations to the local church to support programs for the poor, and establishment of his own charities, including a recreational hall with books and games for workingmen, but no alcohol. But a few of Darwin's well-placed friends, spearheaded by Huxley, lobbied the proper ecclesiastical and parliamentary authorities to secure a public burial in Westminster Abbey, where Darwin lies today, literally at the feet of Isaac Newton.

As a perpetual publicist for the good name of science, Huxley must have relished the prospect that a freethinker who had so discombobulated the most hallowed traditions of Western thought could now lie with kings and conquerors in the most sacred British spot of both political and ecclesiastical authority. But let us be a bit more charitable and grant—even to the combative Huxley, but certainly to the clerics and MPs who made the burial possible—a motivation prompted by a spirit of reconciliation, and by the strong and positive symbol represented by a revolutionary man of science, and at least an agnostic in personal belief, lying in the holiest of Christian holies because he had not feared to seek knowledge and understood that whatever he found could not confute a genuine sense of religion.

Mr. Bridge, the organist of Westminster, composed a funeral anthem for Darwin's interment (a perfectly serviceable piece of music, which I have actively enjoyed under another hat as a choral singer). Bridge chose one of the great biblical wisdom texts, and I cannot imagine a more appropriate set of verses, both for Darwin's ultimate celebration, and for NOMA's theme that a full life—that is, a wise life—requires study and resolution within several magisteria of our complex lives and mentalities.

Happy is the man that findeth wisdom and getteth
    understanding . . .
She is more precious than rubies, and all that thou
    hast cannot be compared unto her.

Length of days is in her right hand; and in her left
hand riches and honor.
Her ways are ways of pleasantness, and all her paths
are peace.

—Modified by Bridge
from Proverbs 3:13–17

A fine statement indeed. I only wish that Mr.
Bridge had added the very next line (Proverbs 3:18)—
the even more famous wisdom text that also happens to
serve as the standard metaphor for evolution as well!

She is a tree of life to them that lay hold upon her.

2

THE PROBLEM
RESOLVED IN
PRINCIPLE

NOMA Defined and Defended

H E COULD CERTAINLY AFFORD THE fees, or simply command the performance by imperial decree, but has any student ever been so blessed in the quality of a private tutor than Alexander the Great, who got several years of undivided attention from Aristotle himself? Now Aristotle preached, as a centerpiece of his philosophy, the concept of a "golden mean," or the resolution of most great issues at a resting point between extremes.

But I wonder how well Aristotle's pupil learned his lessons when I contemplate the two radically different, indeed diametrically opposed, versions of his most famous anecdote. The usual story holds that Alexander, at the height of his military expansion, wept because he had no new worlds to conquer—the dilemma of boredom when "been there, done that" applies to all potential projects. But Plutarch's version, from the first century A.D. and therefore relatively close to the source, features a precisely opposite problem—the dilemma of impotence in a universe too vast to encompass, or even to dent. Plutarch's account also becomes slightly more believable in expressing Aristotle's own doctrine of the eternity of worlds: "Alexander wept when he heard . . . that there was an infinite number of worlds, [saying] 'Do you not think it a matter worthy of lamentation that when there is such a vast multitude of them, we have not yet conquered one?' "

But maybe Alexander understood the golden mean after all, for if we add these extreme stories and divide by two, we may find an intermediate resting place of satisfaction for past achievements, combined with sufficient stimulation for further activity—and therefore no cause for any tears.

I am, of course, only jesting feebly about a symbol chosen to represent the general concept of resolution. Still, I wish to raise a serious point about our usual approach to complex problems, a theme well illustrated by these opposite versions of Alexander's anecdote. Our minds tend to work by dichotomy—that is, by conceptualizing complex issues as "either/or" pairs, dictating a choice of one extreme or the other, with no middle ground (or golden mean) available for any alternative resolution. (I suspect that our apparently unavoidable tendency to dichotomize represents some powerful baggage from an evolutionary past, when limited consciousness could not transcend "on or off," "yes or no," "fight or flee," "move or rest"—and the neurology of simpler brains became wired in accordance with such exigencies. But we must leave this speculative subject for another time and place.)

Thus, when we must make sense of the relationship between two disparate subjects (science and religion in this case)—especially when both seem to raise similar questions at the core of our most vital concerns about life and meaning—we assume that one of two extreme solutions must apply: either science and religion must battle to the death, with one victorious and the other defeated; or else they must represent the same quest and can therefore be fully and smoothly integrated into one grand synthesis.

But both extreme scenarios work by elimination— either the destruction of one by another, or the merger of both into a large and pliant "whole ball of wax" without sharp edges or incisive points. Why not opt instead for a "golden mean" that grants dignity and distinction to *each* subject? We might borrow a paradoxical line from the English essayist G. K. Chesterton, who was not just indulging a national stereotype for dousing anything vibrant and spontaneous with the voice of stolid and restrictive "reason" ("no sex, please, we're

British"), but who epitomized a profound insight about breaking impasses and gaining insight when he stated that "art is limitation; the essence of every picture is the frame."

Consider any of the classically "big" and diffuse "core" questions that have troubled people since the dawn of consciousness: for example, how are humans related to other organisms, and what does this relationship mean? This question contains such richness that no single formulation, and no simple answer, can possibly provide full satisfaction. (All questions of such scope also embody a good deal of "slop" and loose construction, requiring clarification and agreement about intended definitions before any common ground can be sought.)

At this point we must invoke Chesterton's notion of framing and this book's central theme of NOMA, or non-overlapping magisteria. Think of any cliché or standard epigram about distinct items that don't mix— the oil and water, or apples and oranges, of American usage; the chalk and cheese of the corresponding British motto; the two human traditions that cannot join ("and never the twain shall meet") at least until divine power ends the present order of things in Kipling's imperial world ("Till Earth and Sky stand presently at God's great Judgment Seat"). Each domain of inquiry frames its own rules and admissible questions, and sets its own criteria for judgment and resolution. These accepted standards, and the procedures developed for debating and resolving legitimate issues, define the magisterium—or teaching authority—of any given realm. No single magisterium can come close to encompassing all the troubling issues raised by any complex subject, especially one so rich as the meaning of our relationship with other forms of life. Instead of supposing that a single approach can satisfy our full set of concerns ("one size fits all"), we should prepare to visit a picture gallery, where we can commune with several different canvases, each circumscribed by a sturdy frame.

As an example of NOMA applied to a "core issue," let us focus on two distinct frames—that is, two non-overlapping magisteria—surrounding quite different, but equally vital, questions in our search for the meaning of our relationship with other living creatures. On the one hand, we seek information about matters of fact with potential "yes or no" answers (at least in principle; in practice, these answers may be quite difficult to achieve). Some factual questions engage issues of the broadest scale. More than a century ago, for example, the basic formulation of evolutionary theory resolved several problems of this magnitude: Are we related to other organisms by genealogical ties or as items in the ordered scheme of a divine creator? Do humans look so much like apes because we share a recent common ancestor or because creation followed a linear order, with apes representing the step just below us? Other questions, more detailed and more subtle, remain unanswered today: Why does so much of our genetic material (so-called "junk DNA") serve no apparent function? What caused the mass extinctions that have punctuated the history of life? (We pretty well know that an impacting extraterrestrial body triggered the last event 65 million years ago, wiping out dinosaurs and giving mammals a chance, but we have not resolved the causes of the other four major dyings.)

As explained in the Preamble, such questions fall under the magisterium of an institution that we have named "science"—a teaching authority dedicated to using the mental methods and observational techniques validated by success and experience as particularly well suited for describing, and attempting to explain, the factual construction of nature.

But the same subject of our relationship with other organisms also raises a host of questions with an entirely different thrust: Are we worth more than bugs or bacteria because we have evolved a much more complex neurology? Under what conditions (if ever) do we have a right to drive other species to extinction by elimination of their habitats? Do we violate any moral codes when we use genetic technology to place a gene from one creature into the genome of another species? Such questions—and we could fill a long book with just a surface-skimming list—treat the same material of "us and them," but engage different concerns that simply cannot be answered, or even much illuminated, by factual data of any kind. No measure of mental power in humans versus ants will resolve the first question, and no primer on the technology of lateral genetic transfer will provide much help with the last issue.

These questions address moral issues about the value and meaning of life, both in human form and more widely construed. Their fruitful discussion must proceed under a different magisterium, far older than science (at least as a formalized inquiry), and dedicated to a quest for consensus, or at least a clarification of assumptions and criteria, about ethical "ought," rather than a search for any factual "is" about the material construction of the natural world.[1] This magisterium

---

[1] I apologize to colleagues in philosophy and related fields for such an apparently cavalier "brush by" of an old and difficult topic still subject to much discussion, and requiring considerable subtlety and nuancing to capture the ramifying complexities. I recognize that this claim for separation of the factual from the ethical has been controversial (and widely controverted) ever since David Hume drew an explicit distinction between "is" and "ought." (I even once wrote an embarrassingly tendentious undergraduate paper on G. E. Moore's later designation of this issue, in his *Principia Ethica* of 1903, as "the naturalistic fallacy.") I acknowledge the cogency of some classical objections to strict separation—

of ethical discussion and search for meaning includes several disciplines traditionally grouped under the humanities—much of philosophy, and part of literature and history, for example. But human societies have usually centered the discourse of this magisterium upon an institution called "religion" (and manifesting, under this single name, an astonishing diversity of approaches, including all possible beliefs about the nature, or existence for that matter, of divine power; and all possible

particularly the emptiness of asserting an "ought" for behaviors that have been proven physically impossible in the "is" of nature. I also acknowledge that I have no expertise in current details of academic discussion (although I have tried to keep abreast of general developments). Finally, I confess that if an academic outsider made a similarly curt pronouncement about a subtle and troubling issue in my field of evolution or paleontology, I'd be pissed off.

I would, nonetheless, defend my treatment not as a dumbing down, or as disrespect for the complexity of a key subject, but as a principled recognition that most issues of this scope require different treatments at various scales of inquiry. Broad generalizations always include exceptions and nuanced regions of "however" at their borders—without invalidating, or even injuring, the cogency of the major point. (In my business of natural history, we often refer to this phenomenon as the "mouse from Michigan" rule, to honor the expert on taxonomic details who always pipes up from the back of the room to challenge a speaker's claim about a general evolutionary principle: "Yes, but there's a mouse from Michigan that . . .") Among experts, attention properly flows to the exceptions and howevers—for these are the interesting details that fuel scholarship at the highest levels. (For example, my colleagues in evolutionary theory are presently engaged in a healthy debate about whether a limited amount of Lamarckian evolution may be occurring for restricted phenomena in bacteria. Yet the attitudes to freedom of discussion vs. obedience to unchangeable texts or doctrines).

I most emphatically do not argue that ethical people must validate their standards by overt appeals to religion—for we give several names to the moral discourse of this necessary magisterium, and we all know that atheists can live in the most firmly principled manner, while hypocrites can wrap themselves in any flag, including (most prominently) the banners of God and country.

fascination and intensity of this question does not change the well-documented conclusion that Darwinian processes dominate in the general run of evolutionary matters.) But the expert's properly intense focus on wriggles at the border should not challenge or derail our equally valid broad-scale focus on central principles. The distinction of "is" from "ought" ranks as such a central principle, and this little volume has been written (for all intelligent readers, and without compromise or dumbing down) as a broad-scale treatment.

To cite an analogy: At the Arkansas creationism trial (discussed in detail in chapter 3), philosopher Michael Ruse presented the famous Popperian definition of falsifiability as a chief criterion for designating a topic as scientific (with unfalsifiable "creation science" banned by this standard). Judge Overton accepted Ruse's analysis and used this criterion as his main definition of science in reaching his decision to strike down the Arkansas "equal time" law. But falsificationism (like the is-ought distinction, and like Darwinian domination versus a little bacterial Lamarckism) represents a good generality, subject to extensive debate and controversion for several borderland subthemes among professional scholars. Some academic philosophers attacked Ruse for "simplifying" the subtleties of their field, but I would strongly defend his testimony (as did, I believe, the great majority of professional philosophers) as a valid analysis for the appropriate general scale of broad definitions.

But I do reiterate that religion has occupied the center of this magisterium in the traditions of most cultures.

Since every one of us must reach some decisions about the rules we will follow in conducting our own lives (even if we only pledge ourselves to the doctrine of unstinting self-promotion, whatever the cost to other people)—and since I trust that no one can be entirely indifferent to the workings of the world around us (if only to learn enough about the speed of moving vehicles that we don't step into lanes of rapid traffic whenever we wish to cross the street)—all human beings must pay at least rudimentary attention to both magisteria of religion and science, whatever we choose to name these domains of ethical and factual inquiry. Mere existence may be sustained by the minimal concern caricatured above. But real success—at least in the old-fashioned sense of genuine stature—requires serious engagement with the deep and difficult issues of both magisteria. The magisteria will not fuse; so each of us must integrate these distinct components into a coherent view of life. If we succeed, we gain something truly "more precious than rubies," and dignified by one of the most beautiful words in any language: wisdom.

I have advanced two primary claims in designating my conception of the proper relationship between science and religion as NOMA, or non-overlapping magisteria: first, that these two domains hold equal worth and necessary status for any complete human life; and second, that they remain logically distinct and fully separate in styles of inquiry, however much and however tightly we must integrate the insights of both magisteria to build the rich and full view of life traditionally designated as wisdom. Thus, before presenting some examples (in this chapter's more concrete second half) to anchor the generalities of this first section, I must defend these two key claims about NOMA in the face of an evident challenge inherent in the structure of my foregoing argument.

1. Equal status of the magisteria. I am a scientist by profession and a theological skeptic and nonparticipant by confession (as stated on page 8, whatever my sincerely expressed fascination for religion as a subject). Am I truly practicing what I preach about equal and ineluctable status for both magisteria, when one consumes my life, but the other only piques my interest? In particular, how can I defend a professed respect for religion when I seem to denigrate the enterprise by two clear implications of the foregoing discussion? Why shouldn't readers view me as just another arrogant scientist, hypocritically claiming noninterference based on deep respect and affection while actually attempting to demote religion to impotence and inconsequentiality?

As a first implication for potential suspicion, I have

stated that, while every person must formulate a moral theory under the magisterium of ethics and meaning, and while religion anchors this magisterium in most cultural traditions, the chosen pathway need not invoke religion at all, but may ground moral discourse in other disciplines, philosophy for example. If we all must develop a moral code, but may choose to do so without a formal appeal to religion, then how can this subject claim equal importance and dignity with science (which cannot be similarly ignored unless a person truly believes that each step might launch him into outer space rather than force a gravitational return of foot to ground)?

Returning to a previous example, T. H. Huxley reported his distress upon hearing a standard line in the Anglican burial service suggesting that a belief in resurrection serves as a necessary prod for decent behavior during our earthly life:

> As I stood behind the coffin of my little son the other day, with my mind bent on anything but disputation, the officiating minister read, as a part of his duty, the words, "If the dead not rise again, let us eat and drink, for tomorrow we die." I cannot tell you how inexpressibly they shocked me . . . What! because I am face to face with irreparable loss, because I have given back

to the source from whence it came, the cause of a great happiness, still retaining through all my life the blessings which have sprung and will spring from that cause, I am to renounce my manhood, and, howling, grovel in bestiality? Why, the very apes know better, and if you shoot their young, the poor brutes grieve their grief out and do not immediately seek distraction in a gorge.

But note that Huxley here attacks a specific claim within a particular tradition, not the concept of religion itself. When he says, later in the same letter, that "a deep sense of religion" is "compatible with the entire absence of theology," he must have been thinking about this example. A magisterium, after all, is a site for dialogue and debate, not a set of eternal and invariable rules. So Huxley, in these statements, joins a debate within the magisterium of religion about the moral value of good deeds. He surely stands outside the magisterium of science—and even makes claims later recognized as incorrect in his one citation of a supposed fact (about the grieving of apes) to illustrate a position that can only be decided by moral discourse (the greater value of actions based upon consistent principles rather than feared consequences). Huxley, the supposed scourge of God, is evidently quite content to base his rejection of a rote Christian doctrine on a higher principle that he accepts as religious in essential nature. So let us acknowledge the necessity and centrality of dialogue within this magisterium (on vital questions that science cannot touch), and not quibble about the labels. I will accept both Huxley's view and the etymology of the word itself—and construe as fundamentally religious (literally, binding us together) all moral discourse on principles that might activate the ideal of universal fellowship among people.

As a second and more general implication, am I not more subtly denigrating the entire magisterium of ethics and meaning (or whatever name we choose) by implicitly stating that moral questions cannot be answered absolutely, while only a fool would deny the revolution of planets or the evolution of life? On this point we can only return to the principle of apples and oranges—that is, to NOMA itself. This inaccessibility to absolute resolution must be viewed as a logical property of the form of discourse itself, not as a limitation. (The vitality of this magisterium lies largely in the transcendent importance of moral issues, and questions of meaning, for all thinking and feeling people, not in the style of available resolution—based more on compromise and consensus in this magisterium than on factual demonstration, as in the magisterium of science.) One might as well denigrate the magisterium of science because its powerful offspring, technology, can perform such wonders, while all the resources of this great magisterium can hardly cast a flicker of light upon the oldest and simplest ethical questions that have haunted people since the dawn of consciousness.

2. INDEPENDENCE OF THE MAGISTERIA. How can anyone take this vaunted claim for non-overlapping magisteria seriously when the last few centuries of human history can virtually be defined by claims for deep and inherent conflict between these domains—from the evangelist (and former baseball star) Billy Sunday, who stated that any minister believing in evolution must be "a stinking skunk, a hypocrite and a liar" to Disraeli's rather more eloquent appeal:

> The question is this—Is man an ape or an angel? My lord, I am on the side of the angels. I repudiate with indignation and abhorrence the contrary view, which is, I believe, foreign to the conscience of humanity . . . Man is made in the image of his Creator—a source of inspiration and of solace—a source from which only can flow every right principle of morals and every divine truth . . . It is between those two contending interpretations of the nature of man, and their consequences, that society will have to

decide. Their rivalry is at the bottom of all human affairs.

The resolution of this key question will occupy the second half of this book (effectively all of chapters 3 and 4), so I must defer discussion until then. For now, and as a placeholder in the logic of my argument, I will only state that I am trying to analyze the inherent logic of a case, as viewed with some historical distance from the heat of most intense and immediate battle—and that I am not making any claim about the realities of our intellectual and social histories. (I should also reiterate, as stated up front in the beginning of my Preamble, that NOMA represents a long-standing consensus among the great majority of both scientific and religious leaders, not a controversial or idiosyncratic resolution.) In brief, and as a caricature of an epitome for this book's second half, no institution ever gives up turf voluntarily. The magisterium of science is a latecomer in human history. *Faute de mieux,* theology once occupied this realm of factual inquiry as well. We can hardly expect anyone to withdraw from so much territory without a struggle—no matter how just and true the claim may be that such an apparent retreat can only strengthen the discipline.

Finally, how far apart do the magisteria of science and religion stand? Do their frames surround pictures at opposite ends of our mental gallery, with miles of minefields between? If so, why should we even talk about dialogue between such distantly non-overlapping magisteria, and of their necessary integration to infuse a fulfilled life with wisdom?

I hold that this non-overlapping runs to completion only in the important logical sense that standards for legitimate questions, and criteria for resolution, force the magisteria apart on the model of immiscibility—the oil and water of a common metaphorical image. But, like those layers of oil and water once again, the contact between magisteria could not be more intimate and pressing over every square micrometer (or upon every jot and tittle, to use an image from the other magisterium) of contact. Science and religion do not glower at each other from separate frames on opposite walls of the Museum of Mental Arts. Science and religion interdigitate in patterns of complex fingering, and at every fractal scale of self-similarity.

Still, the magisteria do not overlap—but then, neither do spouses fuse in the best of marriages. Any interesting problem, at any scale (hence the fractal claim above, meant more than metaphorically), must call upon the separate contributions of both magisteria for any adequate illumination. The logic of inquiry prevents true fusion, as stated above. The magisterium of science cannot proceed beyond the anthropology of morals—the documentation of what people believe, including such important information as the relative frequency of particular moral values among distinct cultures, the correlation of those values with ecological and economic conditions, and even (potentially) the adaptive value of certain beliefs in specified Darwinian situations—although my intense skepticism about speculative work in this last area has been well aired in other publications. But science can say nothing about the morality of morals. That is, the potential discovery by anthropologists that murder, infanticide, genocide, and xenophobia may have characterized many human societies, may have arisen preferentially in certain social situations, and may even be adaptively beneficial in certain contexts, offers no support whatever for the moral proposition that we ought to behave in such a manner.

Still, only the most fearful and parochial moral philosopher would regard such potential scientific information as useless or uninteresting. Such facts can never validate a moral position, but we surely want to understand the sociology of human behavior, if only to recognize the relative difficulty of instituting various consensuses reached within the magisterium of morals and meaning. To choose a silly example, we had better appreciate the facts of mammalian sexuality, if only to avoid despair if we decide to advocate uncompromising monogamy as the only moral path for human society, and then become confused when our arguments, so forcefully and elegantly crafted, fare so poorly in application.

Similarly, scientists would do well to appreciate the norms of moral discourse, if only to understand why a thoughtful person without expert knowledge about the genetics of heredity might justly challenge an assertion that some particular experiment in the controlled breeding of humans should be done because we now have the technology to proceed, and the results would be interesting within the internal logic of expanding information and explanation.

From Mutt and Jeff to yin and yang, all our cultures, in their full diversity of levels and traditions, include images of the absolutely inseparable but utterly different. Why not add the magisteria of science and religion to this venerable and distinguished list?

NOMA Illustrated

IN ADVOCATING THE NOMA ARGUMENT over many years, I have found that skeptical friends and colleagues do not challenge the logic of the argument—which almost everyone accepts as both intellectually sound and eminently practical in our world of diverse passions—but rather question my claim that most religious and scientific leaders actually do advocate the precepts of NOMA. We all recognize, of course, that many folks and movements hold narrow and aggressively partisan positions, usually linked to an active political agenda, and based on exalting one side while bashing the other. Obviously, extremists of the so-called Christian right, particularly the small segment dedicated to imposing creationism on the science curricula of American public schools, represent the most visible subgroup of these partisans. But I also include, among my own scientific colleagues, some militant atheists whose blinkered concept of religion grasps none of the subtlety or diversity, and equates this entire magisterium with the silly and superstitious beliefs of people who think they have seen a divinely crafted image of the Virgin in the drying patterns of morning dew on the plate-glass windows of some auto showroom in New Jersey.

I believe that we must pursue a primarily political struggle, not an intellectual discourse, with these people. With some exceptions, of course, people who have dedicated the bulk of their energy, and even their life's definition, to such aggressive advocacy at the extremes do not choose to engage in serious and respectful debate. Supporters of NOMA, and all people committed to the defense of honorable differences, will have to remain vigilant and prevail politically.

Even after we put the extremists aside, however, many people still suppose that major religious and scientific leaders must remain at odds (or at least must interact in considerable tension) because these two incompatible fields inevitably struggle for possession of the same ground. If I can therefore show that NOMA enjoys strong and fully explicit support, even from the primary cultural stereotypes of hard-line traditionalism, then the status of NOMA as a sound position of general consensus, established by long struggle among people of goodwill in both magisteria—and not as a funny little off-the-wall suggestion by a few misguided peacemakers on an inevitable battlefield—should emerge into the clearest possible light.

I will therefore discuss two maximally different but equally ringing defenses of NOMA—examples that could not exist if science and religion have been destined to fight for the same disputed territory: first, religion acknowledging the prerogatives of science for the most contentious of all subjects (attitudes of recent popes toward human evolution); and second, science, at the dawn of the modern age, as honorably practiced by professional clergymen (who, by conventional views, should have undermined rather than promulgated such an enterprise).

1. DARWIN AND THE PAPACY. For indefensible reasons of ignorance and stereotypy, people who do not grow up in Roman Catholic traditions tend to view the pope as an archetypal symbol for a dogmatic traditionalism that must, by definition, be hostile to science. Doctrines of infallibility, pronouncements *ex cathedra*, and so forth, combined with extensive trappings of costume and ritual (all formerly, and formally, conducted in incomprehensible Latin), tend to reinforce this stereotype among people who do not really understand their meaning and function.

(For my own appreciation of an institution that

does not always strive to be explicit or revealing, I remain grateful to an English Jesuit who had abandoned a successful business career to undertake the rigors of a long training lasting nearly twenty years, and whom I met by the chance of adjacent seating one night at the Rome opera many years ago. We spent the next two days in intense discussion. He taught me that his church, at its frequent best and in his words, "is one gigantic debating society." Papal pronouncements may debar further official and public disagreement, but the internal dialogue never abates. Consider only the legendary patience and stubbornness of Job [13:15]: "Though he slay me, yet I will trust in him; but I will maintain mine own ways before him.")

Moreover, one defining historical incident—the trial and forced recantation of Galileo in 1633—continues to dominate our cultural landscape as a primary symbol, almost automatically triggered whenever anyone contemplates the relationship of science and Catholicism. The usual version stands so strongly against NOMA, and marks Pope Urban VIII as such a villain, with Galileo as such a martyred hero, that a model of inherent warfare between the magisteria seems inevitable.

The subject deserves volumes rather than the few paragraphs available here, but we must reject the cardboard and anachronistic account that views Galileo as a modern scientist fighting the entrenched dogmatism of a church operating entirely outside its magisterium, and almost ludicrously wrong about the basic fact of cosmology. I would not urge an entirely revisionist reading. The basic facts cannot be gainsaid: Galileo was cruelly treated (forced to recant on his knees, and then placed under the equivalent of house arrest for the remainder of his life), and he was right; his conflict with the Pope did, to cite the best modern work on the subject (*Galileo, Courtier,* by Mario Biagioli, University of Chicago Press, 1993), represent "the clash between two incompatible worldviews," and Urban did defend the traditional geocentric universe as established dogma. But when we begin to appreciate even the tip of the complex iceberg represented by seventeenth-century life at the court of Rome—a world so profoundly different from our own that modern categories and definitions can only plunge us into incomprehension—then we may understand why our current definitions of science and religion map so poorly upon Galileo's ordeal.

As Biagioli shows, Galileo fell victim to a rather conventional form of drama in the princely courts of Europe. Maffeo Barberini had been Galileo's personal friend and a general patron of the arts and sciences. When Barberini became Pope Urban VIII in 1623, Galileo, now nearing sixty years of age, felt that his moment of "now or never" had arrived. The Church had banned the teachings of Copernicus's heliocentric universe as a fact of nature, but had left a conventional door open in permitting the discussion of heterodox cosmologies as purely mathematical hypotheses.

But Galileo moved too fast and too far in an un- necessarily provocative manner. He had lived his life in necessary pursuit of courtly patronage, but now he fell from grace and into a common role of his time and place. In Biagioli's words: "Galileo's career was pro- pelled and then undone by . . . patronage dynamics . . . The dynamics that led to Galileo's troubles were typical of a princely court: they resemble what was known as 'the fall of the favorite.' "

As a prod for questioning our misleading modern categories, ask yourself why a spiritual leader could compel Galileo at all. Why did the great physicist even consent to argue his case before a Church tribunal in Rome? Then remember that no country called Italy ex- isted in the 1630s, and that the Pope held full secular authority over Rome and much surrounding territory. Galileo had to appear before the Inquisition because this body represented "the law of the land," with full power to convict and execute. Moreover, the papal court may have been uniquely volatile among the princely institu- tions that held sway over segments of Europe: times were particularly tough (as the Roman Church faced the expanding might of the Reformation, right in the midst of the devastating Thirty Years' War); the pope held unusual power as both the secular ruler of specific lands and at least the titular spiritual authority over much greater areas; the papal court, almost uniquely, gained no stability from rules of dynastic succession, for new occupants prevailed by election, and could even be recruited from nonaristocratic backgrounds; finally, most popes attained their roles late in life, so "turnover rates" were unusually high and few incumbents reigned long enough to consolidate adequate power.

Now add to this mix a brilliant hothead who had caused trouble before, and who now mocked prior papal directives (or had, at least, been purposely, even outrageously, provocative) by composing his new book as a supposed dialogue between equally matched advocates, and then putting the arguments for a central earth, the official Church position, into the mouth of a character whose cogency fully matched his name—Simplicio. Urban VIII made a really bad move by the proper judgment of later history, but I have no trouble understanding why he felt miffed, if not betrayed—and such feelings did engender predictable consequences in this earlier age of far different sensibilities and accepted procedures.

The power of Galileo's story continues to haunt any issue involving science and the papacy, today as much as ever. I don't know how else to understand the enormous surprise of scientific commentators, and the banner headlines in newspapers throughout the Western world, when

Pope John Paul II recently issued a statement that struck me as entirely unremarkable and fully consistent with long-standing Roman Catholic support for NOMA in general, and for the legitimate claims of human evolution as a subject for study in particular. After all, I knew that the highly conservative Pope Pius XII had defended evolution as a proper inquiry in the encyclical *Humani Generis*, published in 1950, and that he had done so by central and explicit invocation of NOMA—that is, by identifying the study of physical evolution as outside his magisterium, while further distinguishing such Darwinian concepts from a subject often confused with scientific claims but properly lying within the magisterium of religion: namely the origin and constitution of the human soul.

But, on more careful reading and study, I realized that Pope John Paul's statement of 1996 had added an important dimension to Pius's earlier document issued nearly half a century before. The details of this contrast provide my favorite example of NOMA as used and developed by a religious leader not generally viewed as representing a vanguard of conciliation within his magisterium. If NOMA defines the current view of Urban VIII's direct descendant, then we may rejoice in a pervasive and welcome consensus.[2]

[2]The rest of this section on papal views about evolution has been adapted from an essay previously published in *Leonardo's Mountain of Clams and the Diet of Worms* (Crown, 1998).

Pius XII's *Humani Generis* (1950), a highly traditionalist document written by a deeply conservative man, faces all the "isms" and cynicisms that rode the wake of World War II and informed the struggle to rebuild human decency from the ashes of the Holocaust. The encyclical bears the subtitle "concerning some false opinions which threaten to undermine the foundations of Catholic doctrine," and begins with a statement of embattlement:

> Disagreement and error among men on moral and religious matters have always been a cause of profound sorrow to all good men, but above all to the true and loyal sons of the Church, especially today, when we see the principles of Christian culture being attacked on all sides.

Pius lashes out, in turn, at various external enemies of the Church: pantheism, existentialism, dialectical materialism, historicism, and, of course and preeminently, communism. He then notes with sadness that some well-meaning folks within the Church have fallen into a dangerous relativism—"a theological pacifism and egalitarianism, in which all points of view become equally valid"—in order to include those who yearn for the embrace of Christian religion, but do not wish to accept the particularly Catholic magisterium.

Pius first mentions evolution to decry a misuse by overextension among zealous supporters of the anathematized "isms":

> Some imprudently and indiscreetly hold that evolution . . . explains the origin of all things . . . Communists gladly subscribe to this opinion so that, when the souls of men have been deprived of every idea of a personal God, they may the more efficaciously defend and propagate their dialectical materialism.

Pius presents his major statement on evolution near the end of the encyclical, in paragraphs 35 through 37. He accepts the standard account of NOMA and begins by acknowledging that evolution lies in a difficult area where the domains press hard against each other. "It remains for US now to speak about those questions which, although they pertain to the positive sciences, are nevertheless more or less connected with the truths of the Christian faith."[3]

---

[3]Interestingly, the main thrust of these paragraphs does not address evolution in general, but lies in refuting a doctrine that Pius calls "polygenism," or the notion of human ancestry from multiple parents—for he regards such an idea as incompatible with the doctrine of original sin, "which proceeds from a sin actually committed by an individual Adam and which, through generation, is passed on to all and is in

Pius then writes the well-known words that permit Catholics to entertain the evolution of the human body (a factual issue under the magisterium of science), so long as they accept the divine creation and infusion of the soul (a theological notion under the magisterium of religion).

> The Teaching Authority of the Church does not forbid that, in conformity with the present state of human sciences and sacred theology, research and discussions, on the part of men experienced in both fields, take place with regard to the doctrine of evolution, in as far as it inquires into the origin of the human body as coming from pre-existent and living matter—for the Catholic faith obliges us to hold that souls are immediately created by God.

everyone as his own." In this one instance, Pius may be trangressing the NOMA principle—but I cannot judge, for I do not understand the details of Catholic theology and therefore do not know how symbolically such a statement may be read. If Pius is arguing that we cannot entertain a theory about derivation of all modern humans from an ancestral population rather than through an ancestral individual (a potential fact) because such an idea would question the doctrine of original sin (a theological construct), then I would declare him out of line for letting the magisterium of religion dictate a conclusion with the magisterium of science.

I had, up to here, found nothing surprising in *Humani Generis*, and nothing to relieve my puzzlement about the novelty of Pope John Paul's 1996 statement. But I read further and realized that Pope Pius had said more about evolution, something I had never seen quoted, and something that made John Paul's statement most interesting indeed. In short, Pius forcefully proclaimed that while evolution may be legitimate in principle, the theory, in fact, had not been proven and might well be entirely wrong. One gets the strong impression, moreover, that Pius was rooting pretty hard for a verdict of falsity. Continuing directly from the last quotation, he advises us about the proper study of evolution:

> However, this must be done in such a way that the reasons for both opinions, that is, those favorable and those unfavorable to evolution, be weighed and judged with the necessary seriousness, moderation and measure . . . Some, however, rashly transgress this liberty of discussion, when they act as if the origin of the human body from pre-existing and living matter were already completely certain and proved by the facts which have been discovered up to now and by reasoning on those facts, and as if there were nothing in the sources of divine revelation

which demands the greatest moderation and caution in this question.

To summarize, Pius accepts the NOMA principle in permitting Catholics to entertain the hypothesis of evolution for the human body so long as they accept the divine infusion of the soul. But he then offers some (holy) fatherly advice to scientists about the status of evolution as a scientific concept: the idea is not yet proven, and you all need to be especially cautious because evolution raises many troubling issues right on the border of my magisterium. One may read this second theme of advice-giving in two rather different ways: either as a gratuitous incursion into a different magisterium, or as a helpful perspective from an intelligent and concerned outsider.

In any case, this rarely quoted second claim (that evolution remains both unproven and a bit dangerous)—and not the familiar first argument for NOMA (that Catholics may accept the evolution of the body so long as they embrace the creation of the soul)—defines the novelty and the interest of John Paul's recent statement.

John Paul begins by summarizing Pius's older encyclical of 1950, and particularly by reaffirming NOMA—nothing new here, and no cause for extended publicity:

In his encyclical *Humani Generis* (1950), my predecessor Pius XII had already stated that there was no opposition between evolution and the doctrine of the faith about man and his vocation.

The novelty and news value of John Paul's statement lies, rather, in his profound revision of Pius's second and rarely quoted claim that evolution, while conceivable in principle and reconcilable with religion, can cite little persuasive evidence in support, and may well be false. John Paul states—and I can only say amen, and thanks for noticing—that the half-century between Pius surveying the ruins of World War II and his own pontificate heralding the dawn of a new millennium has witnessed such a growth of data, and such a refinement of theory, that evolution can no longer be doubted by people of goodwill and keen intellect:

Pius XII added . . . that this opinion [evolution] should not be adopted as though it were a certain, proven doctrine . . . Today, almost half a century after the publication of the encyclical, new knowledge has led to the recognition of the theory of evolution as more than a hypothesis. It is indeed remarkable that this theory has been progressively accepted by researchers, following a series of discoveries in various fields of knowledge. The convergence, neither sought nor fabricated, of the results of work that was conducted independently is in itself a significant argument in favor of the theory.

In conclusion, Pius had grudgingly admitted evolution as a legitimate hypothesis that he regarded as only tentatively supported and potentially (as he clearly hoped) untrue. John Paul, nearly fifty years later, reaffirms the legitimacy of evolution under the NOMA principle, but then adds that additional data and theory have placed the factuality of evolution beyond reasonable doubt. Sincere Christians may now accept evolution not merely as a plausible possibility, but also as an effectively proven fact. In other words, official Catholic opinion on evolution has moved from "say it ain't so, but we can deal with it if we have to" (Pius's grudging view of 1950) to John Paul's entirely welcoming "it has been proven true; we always celebrate nature's factuality, and we look forward to interesting discussions of theological implications." I happily endorse this turn of events as gospel—literally, good news. I represent the magisterium of science, but I welcome the support of a primary leader from the other major magisterium of our complex lives. And I recall the wisdom of King Solomon: "As cold waters to a thirsty soul, so is good news from a far country" (Proverbs 25:25).

2. The cleric who out-Newtoned Newton. If NOMA did not work, and religion really did demand the suppression of important factual data at key points of contradiction with theological dogma, then how could the ranks of science include so many ordained and devoted clergymen at the highest level of respect and accomplishment—from the thirteenth-century Dominican bishop Albertus Magnus, the teacher of Thomas Aquinas and the most cogent medieval writer on scientific subjects; to Nicholas Steno, who wrote the primary works of seventeenth-century geology and also became a bishop; to Lazzaro Spallanzani, the eighteenth-century Italian physiologist who disproved, by elegant experiments, the last serious arguments for spontaneous generation of life; to the Abbé Breuil, our own century's most famous student of paleolithic cave art?

In the conventional view of warfare between the magisteria, science began its inevitable expansion at religion's expense during the late seventeenth century, a remarkable period known to historians as "*the* scientific revolution." We all honor the primary symbol of the new order, Isaac Newton, whose achievements were captured by his contemporary Alexander Pope in the most incisive of all epitomes:

Nature and nature's laws lay hid in night
God said "Let Newton be," and all was light.

Many people are then surprised to discover—although the great man made no attempt to disguise his commitments—that Newton (along with all other prominent members of his circle) remained an ardent theist. He spent far more time working on his exegeses of the prophecies of Daniel and John, and on his attempt to integrate biblical chronology with the histories of other ancient peoples, than he ever devoted to physics.

Scientists with strong theological commitments have embraced NOMA in several styles—from the argument of "God as clockwinder" generally followed by Newton's contemporaries, to the "bench-top materialism" of most religious scientists today (who hold that "deep" questions about ultimate meanings lie outside the realm of science and under the aegis of religious inquiry, while scientific methods, based on the spatio-temporal invariance of natural law, apply to all potentially resolvable questions about facts of nature). So long as religious beliefs do not dictate specific answers to empirical questions or foreclose the acceptance of documented facts, the most theologically devout scientists should have no trouble pursuing their day jobs with equal zeal.

The first commandment for all versions of NOMA might be summarized by stating: "Thou shalt not mix the magisteria by claiming that God directly ordains

important events in the history of nature by special interference knowable only through revelation and not
accessible to science." In common parlance, we refer to
such special interference as "miracle"—operationally
defined as a unique and temporary suspension of natural law to reorder the facts of nature by divine fiat. (I
know that some people use the word "miracle" in other
senses that may not violate NOMA—but I follow the
classical definition here.) NOMA does impose this
"limitation" on concepts of God, just as NOMA places
equally strong restrictions upon the imperialistic aims of
many scientists (particularly in suppressing claims for
possession of moral truth based on superior understanding of factual truth in any subject).

All consensuses of this sort develop slowly, and from
inchoate beginnings before later distinctions become
clarified and established. In the early days of modern
science, the conceptual need to place miracles outside
this developing magisterium had not been fully articu-
lated, and the issue generated much discussion, eventually resolved as outlined above (with God's direct action
in the creation of living species persisting as a last
stronghold, long after miraculous action has been abandoned for all the rest of nature's factual realm). Ironically, Newton himself held a fairly lenient view on the
admissibility of miracles to scientific discourse. He certainly recognized the explanatory advantages of God's

working within His own established laws, but he regarded as unnecessarily presumptuous any attempts by students of the natural order thus to confine God's range of potential action. If God wished to suspend these laws for a moment of creative interference, then He would do exactly as He wished, and scientists would have to pursue the task of explanation as best they could.

Interestingly, the sharpest opposition to such latitude within the developing magisterium of science, and the strongest argument for defining miracles as strictly outside the compass of scientific inquiry, arose from the most prominent professional cleric within Newton's orbit of leading scientists, the same Reverend Thomas Burnet who graced our first chapter. This irony of a clergyman's firmest support for NOMA, in direct opposition to Newton's looser view, should convince us that the magisteria need not exist in conflict, and that a committed theologian can also operate as an excellent and equally devoted scientist.

Newton, who had just read his friend's *Sacred Theory of the Earth*, wrote to Burnet in January of 1681, stating his praise but also raising a few critiques. In particular, Newton argued that the problem of fitting God's initial creative work into a mere six days might be solved by supposing that the earth then rotated much more slowly, producing "days" of enormous length. Burnet wrote an impassioned letter in immediate response:

Your kindness hath brought upon you the trouble of this long letter, which I could not avoid seeing you have insisted upon . . . the necessity of adhering to Moses his Hexameron as a physical description . . . To show the contrary . . . hath swelled my letter too much. [A hexameron is a period of six days, and Burnet uses the charmingly archaic form of the genitive case, "Moses his Hexameron," where we would now employ an apostrophe and write Moses' Hexameron.]

Burnet himself did not find the days of Genesis troubling because he had long favored an allegorical interpretation of these passages and held, in any case, that the concept of a "day" could not be defined before the sun's creation on the fourth day of the Genesis sequence. But he rejected Newton's exegesis for a different reason: he feared that Newton would not be able to devise a natural explanation for the subsequent speeding up of the earth's rotation to modern days of twenty-four hours—and that his friend would therefore want to invoke a supernatural explanation. Burnet wrote to Newton: "But if the revolutions of the earth were thus slow at first, how came they to be swifter? From natural causes or supernatural?" (Burnet also raised other objections to Newton's reading: those long early days would stretch the lives of the patriarchs even beyond the already problematical 969 of Methuselah; moreover, although animals would have enjoyed the long, sunny hours of daylight, the extended nights might have become unbearable: "If the day was thus long what a doleful night would there be.")

Newton responded directly to Burnet's methodological concerns, for he knew that his friend wished to avoid all arguments based on miracle in science—an issue far more important than the particular matter of early day lengths. He therefore wrote, confirming Burnet's worst fear:

> Where natural causes are at hand God uses them as instruments in his works, but I do not think them alone sufficient for the creation and therefore may be allowed to suppose that amongst other things God gave the earth its motion by such degrees and at such times as was most suitable to the creatures.

Newton also responded to Burnet's worry about those long nights and their impact on early organisms: "And why might not birds and fishes endure one long night as well as those and other animals endure many in Greenland?"

Newton, one of the smartest of men in all our history, surely scored a point over Burnet in his retort about life above the Arctic Circle. Mark one for the polar bears (and another for the little-known penguins at the other end). But I think that we must grant Burnet the superior argument for a methodological claim now regarded as crucial to the definition of science: the status of miracles as necessarily outside this magisterium. The cleric, not the primary icon of modern science, offered a more cogent defense for basic modes of procedure in achieving fruitful answers. Mark one for NOMA.

Coda and Segue

J. S. HALDANE (1860–1936), A GREAT Scottish physiologist and deeply religious man (also the father of J. B. S. Haldane, the even more famous evolutionary biologist who tended to radicalism in politics and atheism in theology), delivered the Gifford Lectures, a series dedicated to exploring the relationships between science and philosophy, at the University of Glasgow in 1927. Haldane devoted his lecture on "the sciences and religion" to the optimal solution of NOMA, and its central implications for religious thinkers on the subject of miracles and explanations of the natural world. Haldane began:

> It is often supposed that the sciences . . . are essentially incompatible with religion. At present, this is a widespread popular belief for which

there seems at first sight to be a substantial basis; and certainly this belief is common among scientific men themselves, although they may say little about it, out of respect for those who do hold sincere religious beliefs and whose lives they admire.

Haldane then locates the major barrier to NOMA in confusion of all forms of religious belief with the particular claim—which does mix the magisteria in contention and would therefore preclude NOMA—that much of material nature has been constructed by miracles inaccessible in principle to scientific study:

To those who believed that religion is dependent on a belief in supernatural intervention it seemed to be dying the death of other superstitions. Yet as a matter of fact religion continued to appeal to men as strongly as before, or perhaps more strongly. . . . I think that [I can] make clear the underlying explanation of this. If my reasoning has been correct, there is no real connection between religion and the belief in supernatural events of any sort or kind.

Finally, Haldane insists that this attitude toward miracles flows from his own deep and active commitment to religion, and not from any protective attitude toward his own magisterium of science:

> I can put my heart into this attempt [to formu-
> late the proper relationship between science and
> religion] because no one can feel more strongly
> than I do that religion is the greatest thing in
> life, and that behind the recognized Churches
> there is an unrecognized Church to which all
> may belong, though supernatural events play no
> part in its creed.

Haldane's argument underlines the toughness of NOMA and provides an apt transition to the second half of this book, where I ask why so many people continue to reject such a humane, sensible, and wonderfully workable solution to the great nonproblem of our times. NOMA is no wimpish, wallpapering, superficial device, acting as a mere diplomatic fiction and smoke screen to make life more convenient by compromise in a world of diverse and contradictory passions. NOMA is a proper and principled solution—based on sound philosophy— to an issue of great historical and emotional weight. NOMA is tough-minded. NOMA forces dialogue and respectful discourse about different primary commit-ments. NOMA does not say "I'm OK, you're OK—so let's just avoid any talk about science and religion."

As such, NOMA imposes requirements that become very difficult for many people. In particular, NOMA does challenge certain particular (and popular) versions of religious belief, even while strongly upholding the general importance of religion. And NOMA does forbid scientific entry into fields where many arrogant scientists love to walk, and yearn to control. For example, if your particular form of religion demands a belief that the earth can only be about ten thousand years old (because you choose to read Genesis as a literal text, whatever such a claim might mean), then you stand in violation of NOMA—for you have tried to impose a dogmatic and idiosyncratic reading of a text upon a factual issue lying within the magisterium of science, and well resolved with a radically different finding of several billion years of antiquity.

The fallacies of such fundamentalist extremism can be easily identified, but what about a more subtle violation of NOMA commonly encountered among people whose concept of God demands a loving deity, personally concerned with the lives of all his creatures—and not just an invisible and imperious clockwinder? Such people often take a further step by insisting that their God mark his existence (and his care) by particular factual imprints upon nature that may run contrary to the findings of science. Now, science has no quarrel whatever with anyone's need or belief in such a personalized concept of divine power, but NOMA does preclude the additional claim that such a God must arrange the facts of nature in a certain set and predetermined way. For example, if you believe that an adequately loving God must show his hand by peppering nature with palpable miracles, or that such a God could only allow evolution to work in a manner contrary to facts of the fossil record (as a story of slow and steady linear progress toward *Homo sapiens*, for example), then a particular, partisan (and minority) view of religion has transgressed into the magisterium of science by dictating conclusions that must remain open to empirical test and potential rejection.

Similarly, to the scientist who thinks that he has gained the right to determine the benefits and uses of a new and socially transforming invention merely because he made the potentiating discovery and knows more than anyone else about the technical details—and who resents the moral concerns of well-informed citizens, especially their insistence upon some role in a dialogue about potential regulation—NOMA answers with equal force that facts of nature cannot determine the moral basis of utility, and that a scientist has no more right to seek such power than his fundamentalist neighbor can muster in trying to become dictator of the age of the earth.

Thus, NOMA works as a taskmaker, not an enabler—

and NOMA therefore cannot expect to sweep toward victorious consensus amid universal smiles, and shouts of hosanna from both sides. But NOMA's success can only be liberating and expansive for all seekers of wisdom.

3

Historical Reasons for Conflict

The Contingent Basis for Intensity

ANDREW DICKSON WHITE (1832–1918), THE first president of Cornell University, also served as the American minister to Russia in the mid-1890s. Soon afterward, in 1896, he published a two-volume work that became one of the most influential books of the *fin de* (last) *siècle*: *A History of the Warfare of Science with Theology in Christendom*. White began his account with a metaphor based on a Russian memory. In early April he looks out from his room above the Neva River in St. Petersburg at a crowd of peasants using their picks to break the ice barrier still damming the river as the spring thaw approaches. The peasants are cutting hundreds of small channels through the ice, so that the swollen river behind may flow gently through, and not burst the dam in a great flood initiated by sudden collapse of the entire barrier:

The waters from thousands of swollen stream-lets above are pressing behind [the ice dam]; wreckage and refuse are piling up against it; every one knows that it must yield. But there is a danger that it may break suddenly, wrenching even the granite quays from their foundations, bringing desolation to a vast population ... The patient mujiks are doing the right thing. The barrier, exposed more and more to the warmth of spring by the scores of channels they are making, will break away gradually, and the river will flow on beneficent and beautiful.

In White's complex metaphor, the flowing river represents human progress, while ice marks the chill imposed by dogmatic theology upon the findings of science. Progress cannot be impeded indefinitely, and if theology does not yield its former control over the proper magisterium of science, then religion, with all its virtues, will die in a cultural or political explosion destructive to all humanity. But if theology—with goodwill, thoughtfully, and step by step—cedes this disputed ground to the rightful occupants of science, then the river of progress can flow gently on, just as the Neva will not flood if the mujiks make enough little channels through the wall of ice.

Interestingly, White did not formulate his thesis about warfare between science and theology primarily to advance the cause of science, but rather to save religion from its own internal enemies. In trying to establish Cornell as a nondenominational university, White had been greatly frustrated by the opposition of local clergy who regarded such a secular institution as the devil's work. He wrote:

> Opposition began at once . . . from the good protestant bishop who proclaimed that all professors should be in holy orders, since to the Church alone was given the command "Go, teach all the nations," to the zealous priest who published a charge that . . . a profoundly Christian scholar had come to Cornell in order to inculcate infidelity . . . from the eminent divine who went from city to city denouncing the "atheistic and pantheistic tendencies" of the proposed education, to the perfervid minister who informed a denominational synod that Agassiz, the last great opponent of Darwin, and a devout theist, was "preaching Darwinism and atheism" in the new institution.

White, who was personally devout and more interested in religion than in science, wrote about his work with Ezra Cornell: "Far from wishing to injure Christianity, we both hoped to promote it; but we did not confound religion with sectarianism." White presented his basic thesis in the introduction to his book:

> In all modern history, interference with science in the supposed interest of religion, no matter how conscientious such interference may have been, has resulted in the direst evils both to religion and to science ... on the other hand, all untrammelled scientific investigation, no matter how dangerous to religion some of its stages may have seemed for the time to be, has invariably resulted in the highest good both of religion and of science.

While we can only applaud White's intentions, his influential model of warfare between two inexorably opposed forces vying for the same turf—a common late-nineteenth-century trope, by the way, and a metaphor strongly (and in this context ironically) promoted by the common cultural reading of Darwin's key phrases about "struggle for existence" and "survival of the fittest"—has generated unfortunate consequences for the perennial discussion of relationships between science and religion. Although White meant only to castigate dogmatic theology—in the interests of promoting true religion, as noted above—his thesis has usually been read in a superficial and self-serving manner as a claim that human progress requires a victory of science over the entire institution of religion.

This unfortunate confusion can also be traced to the second major book in this literary genre, the earlier and equally popular work by the trained physician and avocational historian John William Draper, published in 1874 and titled *History of the Conflict Between Religion and Science*. Draper, far less subtle than White, and far less friendly to religion, also meant "dogmatic and sectarian theology" when he wrote "religion" in his title, but Draper's text can be legitimately read as an attack upon religion, or at least upon a particular religion—for, while he held hope for a supportive relationship between science and Protestantism, Draper strongly promoted the all-too-common prejudice of successful old Americans in his time—a virulent anti-Catholicism directed against the religion of most poor immigrants, the "great unwashed" who threatened to dilute the native stock.

No matter how logical or humane we may regard the model of NOMA, and no matter how false and simplistic we may judge the alternative notion of inherent warfare between science and religion, no one can deny that overt struggle has characterized many prominent cases of historical interaction between these two institutions. How then can NOMA be defended if patterns of actual history speak in such a different voice? I believe that four major reasons—all artifacts of history or consequences of psychology, rather than defendable arguments against a desirable and eminently reachable goal—can explain this anomaly and help us to grasp why such a laudable argument as NOMA continues to face so many obstacles toward acceptance, or even understanding.

1. As stated before, and to oversimplify history a bit (while remaining true to a basic pattern), the human mind cannot help wondering about the nature of things, both for practical reasons of planting and sailing, and for more general motives inspired by our blessed sense of wonder—as in, why does the sun shine, and why is grass green? At earlier periods of most Western cultures, when science did not exist as an explicit enterprise, and when a more unified sense of the nature of things gathered all "why" questions under the rubric of religion, issues with factual resolutions now placed under the magisterium of science fell under the aegis of an enlarged concept of religion.

The caretakers and intellectuals of religion often treated such questions in a manner that we would regard as scientific today—observing and calculating astronomical cycles, for example, in order to develop calendars for both practical and religious reasons (as in the complex determination of movable festivals like

Easter). But, in the absence of scientific knowledge, and often for narrow or dogmatic reasons, other questions now belonging to the magisterium of science frequently received authoritarian ("how do I know, the Bible tells me so") or oracular ("angels did it") answers now deemed contrary to the spirit of NOMA.

If human nature includes such admirable features as our blessed sense of wonder, we are also driven by less estimable propensities realized in such common principles of action as "Don't give up power or turf voluntarily, even if you hold no right to the territory." I don't know that we have to look much deeper in order to understand why history often features warfare, when NOMA should prevail. All professions include dogmatists and powermongers, and such people often gain positions of influence. Religion once held enough secular power to attract more than its share of such people. Many religious intellectuals have always been happy to cede inappropriate territory to the legitimate domain of science, but others, particularly in positions of leadership, chose not to yield an inch, and then played the old hand of dichotomy to brand the developing magisterium of science as a sinister bunch of usurpers under the devil's command—hence the actual and frequent warfare of science, not with religion in the full sense, but with particular embodiments better characterized as dogmatic theology, and contrary to most people's concept of religion, even while sometimes bearing the official label of a particular creed.

2. General principles don't always animate particulars. These realities of history have caused severe clashes between institutions representing science and religion on many specific issues, even though abstract logic and ordinary goodwill should inspire tolerance under NOMA. And if we recognize the intensity of some contests between particular religious leaders and certain scientific conclusions (the case of Galileo, or our modern battles with creationists), just consider the even greater (and often literal) wars of some religious leaders with contrary political forces—all over turf or power, even if publicly defended in terms of doctrine.

To cite just one obvious example, Draper and White—the originators of the standard model of warfare between science and religion—wrote their books with one of the great dramas of nineteenth-century European history firmly in mind: the long conflict between the founders of the state of Italy and one of the most fascinating and enigmatic figures of his time—the originally liberal, but increasingly embittered and reactionary Pope Pio Nono (Pius IX), who still holds the record for papal longevity (reigning from 1846 until his death in 1878).

Early in his regime, and as a consequence of the

revolutions of 1848, Pio Nono had been forced into exile at Gaeta in the kingdom of Naples (the nation of Italy did not yet exist). He returned to power in 1850, and pursued an ever more conservative and confrontational agenda against surrounding political realities for the rest of his pontificate—culminating in the infamous Syllabus of 1864, listing the eighty "principal errors of our times," and effectively declaring war on modern society, especially on science and the concept of religious tolerance. Pio Nono convened the First Vatican Council in 1869, where he maneuvered an overwhelming vote to affirm the doctrine of papal infallibility (John XXIII's Second Vatican Council, opened in 1962, pursued a starkly different nonconfrontational agenda).

The modern nation of Italy had been proclaimed in 1861, and control of Rome and surrounding areas—where the Pope ruled as a secular "king" of real territory as well as a spiritual prince—became an issue that could not be avoided for long. On September 20, 1870, Italian troops entered Rome after symbolic resistance from the papal armed guard. Pio Nono remained in the Vatican (which Italy left under papal control, a situation that still continues) for the rest of his life, bitterly protesting his loss of power, and proclaiming himself a prisoner.

Now, should this history be interpreted as an episode in the warfare between religion and the modern state? Such a reading would make a mockery of history's complexity. First of all, no monolith called "religion" exists. The major struggle in this story occurred *within* the Catholic church, as Pio Nono defeated and purged his own liberal wing. Second, why should we read these events as a tale of religion versus the modern secular state rather than a clash between two political powers, each using the rhetorical tools at its command? So, if a genuine battle, over real territory, between a major religion and a new nation can't be viewed as a war of inherently opposing institutions, why should we accept such a model for the more diffuse, less clearly definable, and generally less contentious dialogues of science and religion? Liberal clergymen of all major faiths have always welcomed and respected science, while many leading scientists remain conventionally devout in their religious beliefs.

3. When scientific conclusions have been denied on grounds explicitly identified as religious by supporters of a contrary view, the subjects involved almost always cut closest to the psychological bone of our deepest hopes and fears—to such questions as "what is man [meaning all of us, despite the language of the King James Bible] that thou art mindful of him?"

To be sure, scientific facts relevant to certain aspects of this question cannot resolve issues about spiritual values or ultimate meanings—subjects under the magisterium of religion. But the factual conclusion that we last shared common ancestry with apelike ancestors some 5 to 8 million years ago does scare the bejesus out of many folks who haven't grappled with NOMA, and who fear that anything other than divine creation *ex nihilo* might rob human life of a special status necessary for personal equanimity in a world of frequent tragedy. One may identify another person's comfort as illogical, but one may not deny the psychological reality of such solace, or even its potential necessity in whatever formula an individual follows to persevere against hardship. Such beliefs about matters of fact will not be surrendered lightly, even if religious faith remains logically immune to the contrary findings of science. And don't forget the comforting answer that Psalm 8 gives to the searing question posed above: "Thou madest him to have dominion over the works of thy hands; thou hast put all things under his feet: all sheep and oxen, yea, and the beasts of the field; the fowl of the air, and the fish of the sea, and whatsoever passeth through the paths of the seas."

4. If science and religion, when properly separated by the NOMA principle, stood far apart and never discussed the same subject again, then our long history of unnecessary and illogical conflict could perhaps be

closed. But, as noted previously (page 65), the two magisteria—to use a pair of organic metaphors—belly right up to each other, and interdigitate in the most intimate and complex manner. Science and religion must ask different, and logically distinct, questions—but their subjects of inquiry are often both identical and maximally meaningful. Science and religion stand watch over different aspects of all our major flashpoints. May they do so in peace and reinforcement—and not like the men who served as cannon fodder in World War I, dug into the trenches of a senseless and apparently interminable conflict, while lobbing bullets and canisters of poison gas at a supposed enemy, who, like any soldier, just wanted to get off the battlefield and on with a potentially productive and rewarding life.

Columbus and the Flat Earth: An Example of the Fallacy of Warfare Between Science and Religion

EVERY SCHOOLCHILD KNOWS—OR AT least knew before contemporary "political correctness" deep-sixed the good admiral full fathom five—the story of brave Christopher Columbus, who discovered America against a nearly universal conviction that he would sail right off the edge of a flat earth instead. This silliest and most flagrantly false of all tales within the venerable genre of "moral lessons for kiddies" provides the best example I know for exposing the harm done by the false model of warfare between science and religion—for we can trace the origin of the myth directly to the formulation of this model by Draper and White. Perhaps the generalities of the preceding section provide enough fuel to carry this particular argument for NOMA, based on proving the falsity of the opposite "warfare" model. But, as an essayist at heart, I believe that the best illustration of a generality lies in a well-chosen and adequately documented "little" example—not in a frontal assault on the abstraction itself (a strategy that can rarely proceed beyond tendentious waffling without the support of interesting details).[1]

We all know that classical scholars established the earth's sphericity. Aristotle's cosmology assumed a spherical planet, and Eratosthenes actually measured the earth's circumference in the third century B.C. The flat-earth myth argues that this knowledge was then lost when ecclesiastical darkness settled over Europe. For a thousand years, almost all scholars held that the earth must be flat—like the floor of a tent, held up by the canopy of the sky, to cite a biblical metaphor read literally. The Renaissance then rediscovered classical notions of sphericity, but proof required the bravery of Columbus and other great explorers who should have sailed off the edge, but (beginning with Magellan's expedition) returned home from the opposite direction after going all the way round.

The inspirational, schoolchild version of the myth centers upon Columbus, who supposedly overcame the calumny of assembled clerics in an epic battle at Salamanca between freedom of thought and religious dog-

[1]Much of the rest of this section comes from a previous essay, "The Late Birth of a Flat Earth," published in *Dinosaur in a Haystack* (Harmony Books, 1995).

matism. Consider this version of the legend from a book for primary-school children written in 1887, soon after the myth's invention (but little different from accounts that I read as a child in the 1950s):

> "But if the world is round," said Columbus, "it is not hell that lies beyond that stormy sea. Over there must lie the eastern strand of Asia, the Cathay of Marco Polo" . . . In the hall of the convent there was assembled the imposing company—shaved monks in gowns . . . cardinals in scarlet robes . . . "You think the earth is round . . . Are you not aware that the holy fathers of the church have condemned this belief . . . This theory of yours looks heretical." Columbus might well quake in his boots at the mention of heresy; for there was that new Inquisition just in fine running order, with its elaborate bone-breaking, flesh-pinching, thumb-screwing, hanging, burning, mangling system for heretics.

(Some of the quotations, and much of the documentation, in this section come from an excellent book by the historian J. B. Russell, *Inventing the Flat Earth,* Praeger, 1991.)

Dramatic to be sure, but entirely fictitious. No

period of "flat earth darkness" ever occurred among scholars (no matter how many uneducated people may have conceptualized our planet in this way, both then and now). Greek knowledge of sphericity never faded, and all major medieval religious scholars accepted the earth's roundness as an established fact of cosmology. Ferdinand and Isabella did refer Columbus's plans to a royal commission headed by Hernando de Talavera, Isabella's confessor and, following the defeat of the Moors, Archbishop of Granada. This commission, composed of both clerical and lay advisers, did meet at Salamanca among other places. They did pose some sharp intellectual objections to Columbus, but no one questioned the earth's roundness. As a major critique, they argued that Columbus could not reach the Indies in his own allotted time, because the earth's circumference was too great. Moreover, his critics were entirely right. Columbus had "cooked" his figures to favor a much smaller earth, and an attainable Indies. Needless to say, he did not and could not reach Asia, and Native Americans are still called Indians as a legacy of his error.

Virtually all major Christian scholars affirmed our planet's roundness. The Venerable Bede referred to the earth as *orbis in medio totius mundi positus* (an orb placed in the center of the universe) in the eighth century A.D. The twelfth-century translations into Latin of many Greek and Arabic works greatly expanded general ap-

preciation of the natural sciences, particularly as-
tronomy, among scholars—and convictions about the
earth's sphericity both spread and strengthened. Roger
Bacon (1220–1292) and Thomas Aquinas (1225–1274)
affirmed roundness via Aristotle and his Arabic commentators, as did the greatest scientists of later medieval
times, including Nicholas Oresme (1320–1382). All
these men held ecclesiastical orders.

So who, then, was arguing for a flat earth, if all
leading scholars believed in roundness? Villains must be
found for any malfeasance, and Russell shows that the
great English philosopher of science William Whewell
first identified major culprits in his *History of the Induc-
tive Sciences*, published in 1837—two far less significant
characters, including the reasonably well known church
father Lactantius (245–325) and the truly obscure Cosmas Indicopleustes, who wrote his *Christian Topography*
in 547–549. Russell comments: "Whewell pointed to
the culprits . . . as evidence of a medieval belief in a flat
earth, and virtually every subsequent historian imitated
him—they could find few other examples."

I own a copy of Lactantius's *Divinae institutiones*
(Divine precepts), published in Lyons in 1541. This
work does indeed include a chapter titled *De antipodibus*
(On the antipodes), ridiculing the notion of a round
earth with all the arguments about upside-down Aus-
tralians, etc., that passed for humor in my fifth-grade

class. Lactantius writes: "Can there be anyone so inept to believe that men exist whose extremities lie above their heads *[quorum vestigia sint superiora quam capita]* . . . that trees can grow downwards, and that rain, and snow, and hail go upwards instead of falling to earth *[pluvias, et nives, et grandinem sursum versus cadere in terram]*?" And Cosmas did champion a literal view of a biblical metaphor—the earth as a flat floor for the rectangular, vaulted arch of the heavens above.

Purveyors of the flat-earth myth could never deny the plain testimony of Bede, Bacon, Aquinas, and others—so they argued that these men acted as rare beacons of brave light in pervasive darkness. But consider the absurdity of such a position. Who formed the orthodoxy representing this consensus of ignorance? Two minor figures named Lactantius and Cosmas Indicopleustes? Bede, Bacon, Aquinas, and their ilk were not brave iconoclasts. They constituted the establishment, and their convictions about the earth's roundness stood as canonical, while Lactantius and colleagues remained marginal.

Where, then, and why, did the myth of medieval belief in a flat earth arise? Russell's historiographic work gives us a good fix on both times and people. None of the great eighteenth-century anticlerical rationalists—not Condillac, Condorcet, Diderot, Gibbon, Hume, or our own Benjamin Franklin—accused early Christian scholars of believing in a flat earth, though these men scarcely veiled their contempt for medieval versions of Christianity. Washington Irving gave the flat-earth story a good boost in his largely fictional history of Columbus, published in 1828—but his version did not take hold. The legend grew during the nineteenth century, but did not enter the crucial domains of schoolboy pap or tour-guide lingo. Russell did an interesting survey of nineteenth-century history texts for secondary schools, and found that very few mentioned the flat-earth myth before 1870, but that almost all texts after 1880 featured the legend. We can therefore pinpoint the invasion of general culture by the flat-earth myth.

Those years also marked the construction of the model of warfare between science and religion as a guiding theme of Western history. Such theories of dichotomous struggle always need whipping boys and legends to advance their claims. Russell argues that the flat-earth myth achieved its canonical status as a primary homily for the triumph of science under this false dichotomization of Western history. How could a better story for the army of science ever be concocted? Religious darkness destroys Greek knowledge and weaves us into a web of fears, based on dogma and opposed to both rationality and experience. Our ancestors therefore lived in anxiety, restricted by clerical irrationality, afraid that any challenge could only provoke a fall off the edge of the earth into eternal damnation. A fit tale for an intended purpose, but entirely false because few medieval Christian scholars ever doubted the earth's sphericity.

In the preceding section, I traced the genesis of the warfare model of science and religion to the influential books of Draper and White. Both authors used the flat-earth myth as a primary example. Draper began by stating his thesis:

> The history of Science is not a mere record of isolated discoveries; it is a narrative of the conflict of two contending powers, the expansive force of the human intellect on one side, and the compressing arising from traditionary faith and human interests on the other . . . Faith is in its nature unchangeable, stationary; Science is in its nature progressive; and eventually a divergence between them, impossible to conceal, must take place.

From the measured tones of this statement, Draper descended into virulent anti-Catholicism and a near proclamation of war:

> Will modern civilization consent to abandon the career of advancement which has given it so

much power and happiness . . . Will it submit to the dictation of a power . . . which kept Europe in a stagnant condition for many centuries, ferociously suppressing by the stake and the sword every attempt at progress; a power that is founded in a cloud of mysteries; that sets itself above reason and common sense; that loudly proclaims the hatred it entertains against liberty of thought and freedom in civil institutions . . .

Then has it in truth come to this, that Roman Christianity and Science are recognized by their respective adherents as being absolutely incompatible; they cannot exist together; one must yield to the other; mankind must make its choice—it cannot have both.

Equally uncompromising statements of war issued from the other side, as in this proclamation from the First Vatican Council:

Let him be anathema . . .
Who shall say that no miracles can be wrought, or that they can never be known with certainty, and that the divine origin of Christianity cannot be proved by them . . .
Who shall say that human sciences ought to be pursued in such a spirit of freedom that one

may be allowed to hold as true their assertions, even when opposed to revealed doctrine.

Who shall say that it may at times come to pass, in the progress of science, that the doctrines set forth by the Church must be taken in another sense than that in which the Church has ever received and yet receives them.

Them's fighting words indeed. But remember that these fulminations from both sides reflect the political realities of a particular time (as discussed on pages 106–107), not the logical necessities of coherent and unchangeable arguments. Pio Nono's stark proclamation rightly angered scientists, but it also brought great sorrow to liberals and supporters of science within the Church. Moreover, as documented in chapter 2 (pages 74–82) for recent papal attitudes toward human evolution, the Catholic Church has since abandoned this confrontational position, born of a specific set of historical circumstances, and has warmly embraced NOMA.

Draper extolled the flat-earth myth as a primary example of religion's constraint and science's progressive power:

The circular visible horizon and its dip at sea, the gradual appearance and disappearance of ships in the offing, cannot fail to incline intelligent sailors to a belief in the globular figure of the earth. The writings of the Mohammedan astronomers and philosophers had given currency to that doctrine throughout Western Europe, but, as might be expected, it was received with disfavor by theologians . . . Traditions and policy forbade [the papal government] to admit any other than the flat figure of the earth, as revealed in the Scriptures.

Russell comments on the success of Draper's work:

The History of the Conflict is of immense importance, because it was the first instance that an influential figure had explicitly declared that science and religion were at war, and it succeeded as few books ever do. It fixed in the educated mind the idea that "science" stood for freedom and progress against the superstition and repression of "religion." Its viewpoint became conventional wisdom.

White's later book also presents Columbus as an apostle of rationalism against theological dogma. Of Cosmas Indicopleustes' flat-earth theory, for example, he wrote: "Some of the foremost men in the Church devoted themselves to buttressing it with new texts and throwing about it new outworks of theological reasoning; the great body of the faithful considered it a direct gift from the Almighty."

Both Draper and White developed their basic model of science versus theology in the context of a seminal and contemporary struggle all too easily viewed in this light—the battle for evolution, specifically for Darwin's secular version based on natural selection. No issue, certainly since Galileo, had so challenged traditional views about the deepest meaning of human life, and therefore so contacted a domain of religious inquiry as well. It would not be an exaggeration to say that the Darwinian revolution directly triggered this influential nineteenth-century conceptualization of Western history as a war between science and religion. White made an explicit connection (quoted on page 101) in his statement about Agassiz (the founder of the museum where I now work, and a visiting lecturer at Cornell). Moreover, the first chapter of his book treats the battle over evolution, while the second begins with the flat-earth myth.

Draper wraps himself even more fully in a Darwinian mantle. The end of his preface designates five great episodes in the history of science's battle with religion: the debasement of classical knowledge and the descent of the Dark Ages; the flowering of science under early Islam; the battle of Galileo with the Catholic

Church; the Reformation (a plus for an anti-Catholic like Draper); and the struggle for Darwinism. Moreover, no one could claim a more compelling personal license for such a view, for Draper had been an unwilling witness—one might even say an instigator—of the single most celebrated incident in the overt struggle between Darwin and divinity. We all have heard the famous story of Bishop Wilberforce and T. H. Huxley duking it out at the British Association meeting in 1860. But how many people know that their verbal pyrotechnics did not form the stated agenda of this meeting, and only arose during free discussion following the formal paper officially set for this session—an address by the same Dr. Draper on the "intellectual development of Europe considered with reference to the views of Mr. Darwin."

This link between struggles over Darwinism and the construction, by Draper and White, of the mythical model of warfare between science and religion—a model that must be debunked for the NOMA principle to prevail—permits a smooth transition to my inevitable discussion of the most potent and current American battle between scientific evidence and claims advanced in the name of religion—the attempt by biblical fundamentalists, now extending over more than seventy contentious years, to ban the teaching of evolution in American public schools, or at least to demand equal time for creationism on a literal biblical time scale (with an earth no more than ten thousand years old) in any classroom that also provides instruction about evolution. If this battle has played a major role in the twentieth-century cultural history of America, and has consumed the unwelcome time of many scientists (including yours truly) in successful political campaigns to preserve the First Amendment and reject the legislatively mandated teaching of palpable nonsense, then how can NOMA be defended as more than a pipe dream in a utopian world?

Defending NOMA from Both Sides Now: The Struggle Against Modern Creationism

CREATIONISM: A DISTINCTIVELY AMERICAN VIOLATION OF NOMA

The myth of Columbus and the flat earth supports NOMA by the negative strategy of showing how the opposite model of warfare between science and religion often invents battles that never occurred, but arise only as forced inferences from the fictional model. Christian scholars never proclaimed a flat earth against the findings of science and the knowledge of antiquity, and Columbus fought no battles with ecclesiastical authorities over this nonissue. Modern creationism, alas, has provoked a real battle, thus supporting NOMA with a positive example of the principle that all apparent struggles between science and religion really arise from violations of NOMA, when a small group allied to one magisterium tries to impose its irrelevant and illegitimate will upon the other's domain. Such genuine historical battles, therefore, do not pit science against religion, and can only represent a power play by zealots formally allied to one side, and trying to impose their idiosyncratic and decidedly minority views upon the magisterium of the other side.

The saga of attempts by creationists to ban the teaching of evolution, or to force their own fundamentalist version of life's history into science curricula of public schools, represents one of the most interesting, distinctive, and persistent episodes in the cultural history of twentieth-century America. The story features a tempestuous beginning, starring two of the great characters of the 1920s, and also a gratifying end in the favorable Supreme Court decision of 1987. The larger struggle, however, has not terminated, but only shifted ground—as creationist zealots find other ways to impose their will and nonsense, now that the Court's defense of the First Amendment precludes their old strategy of enforcing creationism by state legislation!

Please note that I am discussing only a particular historical episode—fundamentalist attempts to impose creationism on public school curricula by legislative fiat—and not all nuances of argument included under the ambiguous term "creationism." Some personal versions of creation fall entirely within the spirit of

NOMA and bear no relationship to this story—the belief, for example, that God works through laws of evolution over the long time scale determined by geology, and that this style of superintendence may be regarded as a mode of creation.

As a matter of fact, not a necessity of logic, the activists of the creationist movement against the teaching of evolution have been young-earth fundamentalists who believe that the Bible must be literally true, that the earth cannot be more than ten thousand years old, and that God created all species, separately and *ex nihilo*, in six days of twenty-four hours. These people then display a form of ultimate hubris (or maybe just simple ignorance) in equating these marginal and long-discredited factual claims with the entire domain of "religion."

I have no quarrel with fundamentalists who believe in teaching their doctrine in homes and churches, and not by forced imposition upon public schools. I am quite sure that they are wrong about the age of the earth and the history of life, and I will be happy to remonstrate with any advocate who maintains an open mind on these questions (not a common commodity within the movement). Lord knows, we have the right to be wrong, even to be stupid, in a democracy! Thus, I have no problem with the largest and most potentially influential of all creationist groups in America, the Jehovah's Witnesses—for they do not try to impose their theological beliefs upon public school science curricula, and they agree with my view that churches and homes are the proper venue for teaching such private and partisan doctrines. In other words, our struggle with creationism is political and specific, not religious at all, and not even intellectual in any genuine sense. (Sorry to be harsh, but young-earth creationism offers nothing of intellectual merit that I have ever been able to discern—but just a hodgepodge of claims properly judged within the magisterium of science, and conclusively disproved more than a century ago.)

Before presenting a capsule history, I would summarize the peculiarities of our contemporary struggle with creationism in two propositions:

1. The forceful and persistent attempt by young-earth creationists to insinuate their partisan and minority theological dogma into the science curricula of American public schools cannot be read, in any legitimate way, as an episode in any supposedly general warfare between science and religion. If the issue must be dichotomized at all, the two sides might be characterized as supporters versus opponents of NOMA; as defenders of the First Amendment for separation of church and state versus theocrats who would incorporate their certainties as official state policy; or, most generally, as defenders of free inquiry and the right of

teachers to present their best understanding of subjects in the light of their professional training versus the setting of curricula by local sensibilities or beliefs (or just by those who make the most noise or gain transitory power), whatever the state of natural knowledge, or the expertise of teachers.

In any case, however we choose to parse this controversy, the two sides cannot be labeled as science and religion on the most basic criterion of empirical evidence. For the great majority of professional clergy and religious scholars stand *on the same side* with the great majority of scientists—as defenders of NOMA and the First Amendment, and against the imposition of any specific theological doctrine, especially such a partisan and minority view, upon the science curricula of public schools. For example, the long list of official plaintiffs who successfully challenged the Arkansas creationism statute in 1981 included some scientists and educators, but even more ordained clergy of all major faiths, and scholars of religion.

2. This controversy is as locally and distinctively American as apple pie and Uncle Sam. No other Western nation faces such an incubus as a serious political movement (rather than a few powerless cranks at the fringes). The movement to impose creationism upon public school science curricula arises from a set

of distinctively American contrasts, or generalities expressed in a peculiarly American context: North versus
South, urban versus rural, rich versus poor, local or
state control versus federal standards. Moreover, young-
earth creationism can be favored only by so-called fundamentalists who accept the Bible as literally true in
every word—a marginal belief among all major Western religions these days, and a doctrine only well developed within the distinctively American context of
Protestant church pluralism. Such a fundamentalist perspective would make no sense in any predominantly
Catholic nation, where no tradition for reading the
Bible literally (or much at all, for that matter) has ever
existed. Jewish traditions, even among the orthodox,
may revere the Torah as the absolutely accurate word of
God, where neither one jot nor tittle of text can ever
be altered, but few scholars would ever think of inter-
preting this unchangeable text literally.[2]

[2]I am no biblical expert or exegete, and cannot engage this issue in any
serious manner. But I must say that I simply don't understand what
reading the Bible "literally" can mean, since the text, cobbled together
from so many sources, contains frequent and inevitable contradictions.
These variant readings pose no problem to the vast majority of religious people who view the Bible as an inspired document full of
moral truth, and not as an accurate chronicle of human history or a
perfect account of nature's factuality. For the most obvious example,
how can "literalists" reconcile the plainly different creation stories of
Genesis I and II, which, according to all biblical scholars I have ever
consulted, clearly derive from different sources. In the more familiar

Protestantism has always stressed personal Bible study, and justification by faith, rather than through saints or the interpretations of priests—and literalism becomes conceivable under these practices. But, again, the vast majority of modern Protestants would not choose to read their sacred texts in such a dogmatic and uncompromising manner—particularly in European nations with a limited diversity of mainly liberal styles. But American Protestantism has diversified into a

Genesis I, God creates sequentially in six days, moving from light to the division of waters and firmament, to land and plants, to the sun and moon, and finally to animal life of increasing complexity. On the sixth day he creates human beings, both male and female together: "So God created man in his own image, in the image of God created he him; male and female created he them." In Genesis II, God creates the earth and heavens and then makes a man "of the dust of the ground." He then creates plants and animals, bringing all the beasts to Adam and granting his first man naming rights. But Adam is lonely, so God creates a female companion from one of his ribs: "And the Lord God caused a deep sleep to fall upon Adam, and he slept; and he took one of his ribs, and closed up the flesh thereof; and the rib, which the Lord God had taken from man, made he the woman, and brought her unto the man. And Adam said, This is now bone of my bones, and flesh of my flesh: she shall be called Woman." Our traditional reading conflates these two stories, taking the basic sequence, with humans last, from Genesis I, but borrowing the rib scenario for the subsequent creation of Eve from Genesis II. I often surprise people by pointing out this contradiction and conflation (for even highly devout people don't always study the Bible much these days). They think that I must be nuts, or hallucinating, so I just tell them to check it out (at least most homes still have the basic data, no matter how otherwise bookless!)—and they get mighty surprised. Always be wary of what you think you know best.

uniquely rich range of sects, spanning the full gamut of conceivable forms of worship and belief. The vast majority, of course, pursue the same allegorical and spiritual style of reading as their Catholic and Jewish neighbors, but a few groups—mostly Southern, rural, and poor, to cite the distinctive dichotomies mentioned above—have dug in against all "modernism" with a literalist reading not subject to change, or even argument: "Gimme that old-time religion. It was good enough for grandpa, and it's good enough for me." (Through personal ignorance, I am not considering here the traditions of Islam and non-Western religions.)

To cite just one example of fundamentalism's distinctly American base, and of the puzzlement that creationism evokes in the rest of the religious world, I once stayed at the Casa del Clerico in Rome, a hotel maintained by the Vatican, mostly for itinerant priests. One day in the lunchroom, a group of French and Italian Jesuits called me over. They belonged to a group of practicing scientists, visiting Rome for a convention on science and the Church. They had been reading about the growth of "scientific creationism" in America, and were deeply confused. They thought that evolution had been adequately proven, and certainly posed no challenge to religion in any case (both by their own reasoning and by papal pronouncement, as discussed on pages 74–82). So what, they asked, was going on *chez moi*?

Had good scientific arguments for young-earth creationism really been developed, and by lay fundamentalists rather than professional scientists? A wonderful
polyglot of conversation ensued for the next half hour
in all three languages. I told them that no new (or any)
good arguments existed, and that the issues were both
entirely political and uniquely American. They left sat-
isfied, and perhaps with a better sense of the conun-
drum that America represents to the rest of the world.

TROUBLE IN OUR OWN HOUSE: A BRIEF LEGAL SURVEY FROM SCOPES TO SCALIA

The fundamentalist movement may be as old as
America, and its opposition to teaching evolution must
be as old as Darwin. But this marginal, politically disen-
franchised, and largely regional movement could muster
no clout to press a legislative agenda until one of the
great figures of American history, William Jennings
Bryan (much more on him later; see pages 150–170),
decided to make his last hurrah on this issue. Bryan gave
the creationist movement both influence and contacts.
In the early 1920s, several Southern states passed flat-
out anti-evolution statutes. The Tennessee law, for

example, declared it a crime to teach that "man had descended from a lower order of animals."

American liberals, including many clergymen, were embarrassed and caught off guard by the quick (if local) successes of this movement. In a challenge to the constitutionality of these statutes, the American Civil Liberties Union instigated the famous Scopes trial in Dayton, Tennessee, in 1925. John Scopes, a young freethinker, but quite popular among his largely fundamentalist students, worked as the physics teacher and track coach of the local high school. He had substituted for the fundamentalist biology teacher during an illness, and had assigned the chapters on evolution from the class textbook, *A Civic Biology,* by George William Hunter. Scopes consented to be the guinea pig or stalking horse (choose your zoological metaphor) for a legal challenge to the constitutionality of Tennessee's recently enacted anti-evolution law—and the rest is history, largely filtered and distorted, for most Americans, through the fictionalized account of a wonderful play, *Inherit the Wind,* written in 1955 by Jerome Lawrence and Robert Edwin Lee, and performed by some of America's best actors in several versions. (I had the great privilege, as a teenager, to see Paul Muni at the end of his career playing Clarence Darrow in the original Broadway production, with an equally impressive Ed Begley as William Jennings Bryan. Two film versions featured similar talent, with Spencer Tracy as Darrow and Fredric March as Bryan in the first, and Kirk Douglas as Darrow with Jason Robards as Bryan in the later remake for television.)

Contrary to the play, Scopes was not persecuted by Bible-thumpers, and never spent a second in jail. The trial did have its epic moments—particularly when Bryan, in his major speech, virtually denied that humans were mammals; and, in the most famous episode, when Judge Raulston convened his court on the lawn (for temperatures had risen into triple digits and cracks had developed in the ceiling on the floor below the crowded courtroom), and allowed Darrow to put Bryan on the stand as a witness for the defense. But the usual reading of the trial as an epic struggle between benighted Yahooism and resplendent virtue simply cannot suffice—however strongly this impression has been fostered both by *Inherit the Wind* and by the famous reporting of H. L. Mencken, who attended the trial and, to say the least, professed little respect for Bryan, whom he called "a tinpot pope in the Coca-Cola belt."

Scopes was recruited for a particular job—both by the ACLU and by Dayton fundamentalists, who saw the trial as an otherwise unobtainable opportunity to put their little town "on the map"—and not proactively persecuted in any way. The ACLU wanted a quick process and a sure conviction, not a media circus. (The

Scopes trial initiated live broadcast by radio, and might therefore be designated as the inception of a trajectory leading to O.J. Simpson and other extravaganzas of arguable merit.) The local judge held no power to determine the constitutionality of the statute, and the ACLU therefore sought an unproblematical conviction, designated for appeal to a higher court. They may have loved Clarence Darrow as a personality, but they sure as hell didn't want him in Dayton. However, when Bryan announced that he would appear for the state of Tennessee to rout Satan from Dayton, the die was cast, and Darrow's counteroffer could hardly be refused.

The basic facts have been well reported, but the outcome has almost always been misunderstood. Darrow did bring several eminent scientists to testify, and the judge did refuse to let them take the stand. In so deciding, he was not playing the country bumpkin, but making a proper ruling that his court could judge Scopes's guilt or innocence only under the given statute— and Scopes was guilty as charged—not the legitimacy or constitutionality of the law itself. Testimony by experts about the validity or importance of evolution therefore became irrelevant. In this context, historians have never understood why Judge Raulston then allowed Bryan to testify as an expert for the other side. But this most famous episode has also been misread. First of all, the judge later struck the entire testimony from the record. Second, Darrow may have come out slightly ahead, but Bryan parried fairly well, and certainly didn't embarrass himself. The most celebrated moment—when Darrow supposedly forced Bryan to admit that the days of creation might have spanned more than twenty-four hours—represented Bryan's free-will statement about his own and well-known personal beliefs (he had never been a strict biblical literalist), not a fatal inconsistency, exposed by Darrow's relentless questioning.

To correct the other most famous incident of the trial, Bryan did indeed drop dead of heart failure in Dayton—not dramatically on the courtroom floor (as fiction requires for maximal effect), but rather a week later, after stuffing himself at a church dinner. However, the most serious misunderstanding lies with the verdict itself, and the subsequent history of creationism. *Inherit the Wind* presents a tale of free inquiry triumphant over dogmatism. As an exercise in public relations, the Scopes trial may be read as a victory for our side. But the legal consequences could hardly have been more disastrous. Scopes was, of course, convicted—no surprises there. But the case was subsequently declared moot— and therefore unappealable—by the judge's error of fining Scopes one hundred dollars (as the creationism statute specified), whereas Tennessee law required that all fines over fifty dollars be set by the jury. (Perhaps sleepy little towns like Dayton never fined anyone more than fifty bucks for anything, and the judge had simply forgotten this detail of unapplied law.) In any case, this error provides a good argument against using "outside agitators" like Darrow as sole representatives in local trials. The fancy plaintiff's team, lead by Darrow and New York lawyer Dudley Field Malone, included no one with enough local knowledge to challenge the judge and assure proper procedure.

Thus, Scopes's conviction was overturned on a technicality—an outcome that has usually been depicted as a victory, but was actually a bitter procedural defeat that stalled the real purpose of the entire enterprise: to test the law's constitutionality. In order to reach the appropriate higher court, the entire process would have to start again, with a retrial of Scopes. But history could not be rolled back, for Bryan was dead, and Scopes, now enrolled as a graduate student in geology at the University of Chicago, had no desire to revisit this part of his life. (Scopes, a splendidly modest and honorable man, became a successful oil geologist in Shreveport, Louisiana. He never sought any profit from what he recognized as his accidental and transitory fame, and he never wavered in defending freedom of inquiry and the rights of teachers.)

So the Tennessee law (and similar statutes in other states) remained on the books—not actively enforced, to be sure, but ever-present as a weapon against the proper teaching of biology. Textbook publishers, the most cowardly arm of the printing industry, generally succumbed and either left evolution out or relegated the subject to a small chapter at the back of the book. I own a copy of the text that I used in 1956 at a public high school in New York City, a liberal constituency with no compunction about teaching evolution. This text, *Modern Biology,* by Moon, Mann, and Otto, dominated the market and taught more than half of America's high school students. Evolution occupies only 18 of the book's 662 pages—as chapter 58 out of 60. (Many readers, remembering the realities of high school, will immediately know that most classes never got to this chapter at all.) Moreover, the text never mentions the dreaded "E" word, and refers to Darwin's theory as "the hypothesis of racial development." But the first edition of this textbook, published in 1921, before the Scopes trial, featured Darwin on the frontispiece (my 1956 version substitutes a crowd of industrious beavers for the most celebrated of all naturalists), and includes several chapters treating evolution as both a proven fact and the primary organizing theme for all biological sciences.

This sorry situation persisted until 1968, when Susan Epperson, a courageous teacher from Arkansas, challenged a similar statute in the Supreme Court—and won the long-sought verdict of unconstitutionality on obvious First Amendment grounds. (A lovely woman approached me after a talk in Denver last year. She thanked me for my work in fighting creationism and then introduced herself as Susan Epperson. She had attended my lecture with her daughter, who, as a graduate student in evolutionary biology, had reaped the fruits of her mother's courage. I could only reply that the major thrust of thanks must flow in the other direction.)

But nothing can stop a true believer. The creationists regrouped, and came back fighting with a new strategy designed to circumvent constitutional problems. They had always honorably identified their alternative system as explicitly theological, and doctrinally based in a literal reading of the Bible. But now they expurgated their texts, inventing the oxymoronic concept of "creation science." Religion, it seems, and contrary to all previous pronouncements, has no bearing upon the subject at all. The latest discoveries of pure science now reveal a factual world that just happens to correlate perfectly with the literal pronouncements of the Book of Genesis. If virtually all professionally trained scientists regard such a view as nonsensical, and based on either pure ignorance or outright prevarication, then we can only conclude that credentialed members of this discipline cannot recognize the cutting edge of their own subject. In such a circumstance, legislative intervention becomes necessary. And besides, the creationists continued, we're not asking schools to ban evolution anymore (that argument went down the tubes with the Epperson decision). Now we are only demanding "equal time" for "creation science" in any classroom that also teaches evolution. (Of course, if they decide not to teach evolution at all . . . well, then . . .)

However ludicrous such an argument might be, and however obviously self-serving as a strategy to cloak a real aim (the imposition of fundamentalist theological doctrine) in new language that might pass constitutional muster, two states actually did pass nearly identical "equal time" laws in the late 1970s—Arkansas and Louisiana. A consortium of the ACLU and many professional organizations, both scientific and religious, challenged the Arkansas statute in a trial labeled by the press (not inappropriately) as "Scopes II," before Federal Judge William R. Overton in Little Rock during December 1981. Judge Overton, in a beautifully crafted decision (explaining the essence of science, and the proper role of religion, so well that *Science*, our leading professional journal, published the text verbatim), found the Arkansas equal-time law unconstitutional in January 1982.

The state of Arkansas, now back under the liberal leadership of Bill Clinton, decided not to appeal.

Another federal judge then voided the nearly identical Louisiana law by summary judgment, stating that the case had been conclusively made in Arkansas. Louisiana, however, did appeal to the U.S. Supreme Court in *Edwards v. Aguillard*, where, in 1987, we won a strong and final victory by a seven-to-two majority, with (predictably) Rehnquist and Scalia in opposition (Thomas, a probable third vote today, had not yet joined the court).

I testified at the Arkansas trial as one of six "expert witnesses" in biology, philosophy of science, and theology—with my direct examination centered upon creationist distortion of scientific work on the length of geological time and the proof of evolutionary transformation in the fossil record, and my cross-examination fairly perfunctory. (The attorney general of Arkansas, compelled by the ethics of his profession to defend a law that he evidently deemed both silly and embarrassing to his state, did a competent job, but just didn't have his heart in the enterprise.)

As a group, by the way, we did not try to prove evolution in our testimony. Courtrooms are scarcely the appropriate venue for adjudicating such issues under the magisterium of science. We confined our efforts to the only legal issue before us: to proving, by an analysis of their texts and other activities, that "creation science" is nothing but a smoke screen, a meaningless and oxymoronic phrase invented as sheep's clothing for the old wolf of Genesis literalism, already identified in the Epperson case as a partisan theological doctrine, not a scientific concept at all—and clearly in violation of First Amendment guarantees for separation of church and state if imposed by legislative order upon the science curricula of public schools.

I can't claim that the trial represented any acme of tension in my life. The outcome seemed scarcely in doubt, and we held our victory party on the second day of a two-week trial. But cynicism does not run strongly in my temperament—and I expect that when I am ready to intone my *Nunc Dimittis*, or rather my *Sh'ma Yisroel*, I will list among my sources of pride the fact that I joined a group of scholars to present the only testimony ever provided by expert witnesses before a court of law during this interesting episode of American cultural history—the legal battle over creationism that raged from Scopes in 1925 to *Edwards v. Aguillard* in 1987. (Judge Raulston did not allow Darrow's experts to testify at the Scopes trial, and the Louisiana law was dismissed by summary judgment and never tried; live arguments before the Supreme Court last only for an hour, and include no witnesses.) It was, for me, a great joy and privilege to play a tiny role in a historical tale that featured such giant figures as Bryan and Darrow.

The Arkansas trial may have been a no-brainer, but many anecdotes, both comic and serious, still strike me as illuminating or instructive. In the former category, I may cite my two favorite moments of the trial. First, I remember the testimony of a second-grade teacher who described an exercise he uses to convey the immense age of the earth to his students: he stretches a string across his classroom, and then places the children at appropriate points to mark the origin of life, the death of dinosaurs, and human beginnings right next to the wall at the string's end. In cross-examination, the assistant attorney general asked a question that he later regretted: What would you do under the equal-time law if you had to present the alternative view that the earth is only ten thousand years old? "I guess I'd have to get a short string," the teacher replied. The courtroom burst into laughter, evidently all motivated by the same image that had immediately popped into my mind: the thought of twenty earnest second-graders all scrunched up along one millimeter of string.

In a second key moment, the creationist side understood so little about the subject of evolution that they brought, all the way from Sri Lanka, a fine scientist named Chandra Wickramasinghe, who happens to disagree with Darwinian theory (but who is not an anti-evolutionist, and certainly not a young-earth creationist—a set of distinctions that seemed lost on intellectual leaders of this side). Their lawyer asked him, "What do you think of Darwin's theory?" and

Wickramasinghe replied, in the crisp English of his native land, "Nonsense." In cross-examination, our lawyer asked him: "And what do you think of the idea that the earth is only ten thousand years old?" "Worse nonsense," he tersely replied.

On the plane back home, I got up to stretch my legs (all right, I was going to take a pee), and a familiar-looking man, sitting in an aisle seat of the coach section, stopped me and said in the local accent, "Mr. Gould, I wanna thank you for comin' on down here and heppin' us out with this little problem." "Glad to do it," I replied, "but what's your particular interest in the case? Are you a scientist?" He chuckled and denied the suggestion. "Are you a businessman?" I continued. "Oh no," he finally replied, "I used to be the governor. I'd have vetoed that bill." I had been talking with Bill Clinton. In an odd contingency of history that allowed this drama to proceed to its end at the Supreme Court, Clinton had become a bit too complacent as boy-wonder governor, and had not campaigned hard enough to win reelection in 1980—a mistake that he never made again, right up to the presidency. The creationism bill, which he would surely have vetoed, passed during his interregnum, and was signed by a more conservative governor.

But such humor served only as balance for the serious and poignant moments of the trial—none so moving as the dignity of committed teachers who testified that they could not practice their profession honorably if the law were upheld. One teacher pointed to a passage in his chemistry text that attributed great age to fossil fuels. Since the Arkansas act specifically included "a relatively recent age of the earth" among the definitions of creation science requiring "balanced treatment," this passage would have to be changed. The teacher claimed that he did not know how to make such an alteration. Why not? retorted the assistant attorney general in his cross-examination. You only need to insert a simple sentence: "Some scientists, however, believe that fossil fuels are relatively young." Then, in the most impressive statement of the entire trial, the teacher responded: I could, he argued, insert such a sentence in mechanical compliance with the act. But I cannot, as a conscientious teacher, do so. For "balanced treatment" must mean "equal dignity," and I would therefore have to justify the insertion. And this I cannot do, for I have heard no valid arguments that would support such a position.

Another teacher spoke of similar dilemmas in providing balanced treatment in a conscientious rather than a mechanical way. What then, he was asked, would he do if the law were upheld? He looked up and said, in a calm and dignified voice: It would be my tendency not to comply. I am not a revolutionary or a martyr, but

I have responsibilities to my students, and I cannot forgo them.

And now, led back by this serious note, I realize that I have been a bit too sanguine during this little trip down memory lane. Yes, we won a narrow and specific victory after sixty years of contention: creationists can no longer hope to realize their aims by official legislation. But these well-funded and committed zealots will not therefore surrender. Instead they have changed their tactics, often to effective strategies that cannot be legally curtailed. They continue to pressure textbook publishers for deletion or weakening of chapters about evolution. (But we can also fight back—and have done so effectively in several parts of the country—by urging school boards to reject textbooks that lack adequate coverage of this most fundamental topic in the biological sciences.) They agitate before local school boards, or run their own candidates in elections that rarely inspire large turnouts, and can therefore be controlled by committed minorities who know their own voters and get them to the polls. (But scientists are also parents, and "all politics is local," as my own former congressman from Cambridge, MA, used to say.)

Above all—in an effective tactic far more difficult to combat because it works so insidiously and invisibly—they can simply agitate in vociferous and even mildly threatening ways. Most of us, including most teachers, are not particularly courageous, and do not choose to become martyrs. Who wants trouble? If little Billy tells his parents that I'm teaching evolution, and they then cause a predictable and enormous public fuss (particularly in parts of America where creationism is strong and indigenous) . . . well, then, what happens to me, my family, and my job? So maybe I just won't teach evolution this year. What the hell. Who needs such a mess?

Which leads me to reiterate an obvious and final point: We misidentify the protagonists of this battle in the worst possible way when we depict evolution versus creationism as a major skirmish in a general war between science and religion. Almost all scientists and almost all religious leaders have joined forces *on the same side*—against the creationists. And the chief theme of this book provides the common currency of agreement—NOMA, and the call for respectful and supportive dialogue between two distinct magisteria, each inhabiting a major mansion of human life, and each operating best by shoring up its own home while admiring the other guy's domicile and enjoying a warm friendship filled with illuminating visits and discussions.

Creationists do not represent the magisterium of religion. They zealously promote a particular theological doctrine—an intellectually marginal and demographically minority view of religion that they long to impose upon the entire world. And the teachers of Arkansas represent far more than "science." They stand for toleration, professional competence, freedom of inquiry, and support for the Constitution of the United States— a worthy set of goals shared by the vast majority of professional scientists and theologians in modern America. The enemy is not religion but dogmatism and intolerance, a tradition as old as humankind, and impossible to extinguish without eternal vigilance, which is, as a famous epigram proclaims, the price of liberty. We may laugh at a marginal movement like young-earth creationism, but only at our peril—for history features the principle that risible stalking horses, if unchecked at the starting gate, often grow into powerful champions of darkness. Let us give the last word to Clarence Darrow, who stated in his summation at the Scopes trial in 1925:

> If today you can take a thing like evolution and make it a crime to teach it in the public schools, tomorrow you can make it a crime to teach it in the private schools and next year you can make it a crime to teach it to the hustings or in the church. At the next session you may ban books and the newspapers ... Ignorance and fanaticism are ever busy and need feeding. Always

feeding and gloating for more. Today it is the public school teachers; tomorrow the private. The next day the preachers and the lecturers, the magazines, the books, the newspapers. After a while, Your Honor, it is the setting of man against man and creed against creed until with flying banners and beating drums we are marching backward to the glorious ages of the sixteenth century when bigots lighted fagots to burn the men who dared to bring any intelligence and enlightenment and culture to the human mind.

THE PASSION AND COMPASSION OF WILLIAM JENNINGS BRYAN: THE OTHER SIDE OF NOMA

The usual, and heroic, version of evolution versus creation in twentieth-century America stops here, with a chronicle of legal success, some dire warnings about the need for future diligence, and a reaffirmation of intellectual principles. But I must go on, for an important chapter from the other side, a tale rarely told and little known, demands attention in a book dedicated to the principle of NOMA.

The usual view of William Jennings Bryan[3]—three-time loser as a presidential candidate, and preeminent windbag as an orator—provides an easy target for our ridicule, particularly for those of us who represent what might be called "the Northeastern intellectual establishment," and have never fathomed the very different traditions of Midwestern populism, as represented by Bryan, also known as "The Great Commoner." Just consider the unsparing ridicule of H. L. Mencken, who observed Bryan in action at the Scopes trial and wrote:

> Once he had one leg in the White House and the nation trembled under his roars. Now he is a tinpot pope in the Coca-Cola belt and a brother to the forlorn pastors who belabor halfwits in galvanized iron tabernacles behind the railroad yards . . . It is a tragedy, indeed, to begin life as a hero and to end it as a buffoon.

Mencken's harsh judgment underscores a striking paradox. Bryan spent most of his career as a courageous reformer, not as an addlepated Yahoo. How, then, could

[3]Much of the material for this section comes from my essay "William Jennings Bryan's last campaign," published in *Bully for Brontosaurus* (W.W. Norton, 1991).

this man, America's greatest populist, become, late in life, her arch reactionary?

For it was Bryan who, just one year beyond the minimum age of thirty-five, won the Democratic presidential nomination in 1896 with his populist rallying cry for abolition of the gold standard: "You shall not press down upon the brow of labor this crown of thorns. You shall not crucify mankind upon a cross of gold." Bryan who ran twice more, and lost in noble campaigns for reform, particularly for Philippine independence and against American imperialism. Bryan, the pacifist who resigned as Wilson's secretary of state because he sought a more rigid neutrality in the First World War. Bryan who stood at the forefront of most progressive victories in his time: women's suffrage, the direct election of senators, the graduated income tax (no one loves it, but can you think of a fairer way?). How could this man have then joined forces with the cult of biblical literalism in an effort to purge religion of all liberality, and to stifle the same free thought that he had advocated in so many other contexts?

This paradox still intrudes upon us because Bryan forged a living legacy (as the preceding section documented), not merely an issue lost in the mists of history. For without Bryan, there never would have been antievolution laws, never a Scopes trial, never a resurgence in our day, and never a Supreme Court decision. Every one of Bryan's progressive triumphs would have occurred without him. He fought mightily and helped powerfully; nevertheless, women would be voting today and we would be paying income tax if he had never been born. But the legislative attempt to curb evolution was his baby, and he pursued it with all his legendary demoniac fury. No one else in the ill-organized fundamentalist movement had the inclination, and surely no one else had the legal skill or political clout.

This apparent paradox of shifting allegiance forms a recurring theme in literature about Bryan. His biography in the *Encyclopaedia Britannica*, for example, holds that the Scopes trial "proved to be inconsistent with many progressive causes he had championed for so long."

Two major resolutions have been proposed. The first, clearly the majority view, holds that Bryan's last battle was, indeed, inconsistent with all the populist campaigning that had gone before. Who ever said that a man must maintain an unchanging ideology throughout adulthood; and what tale of human psychology could be more familiar than the transition from Young Turk to Old Fart? Most biographies treat the Scopes trial as an inconsistent embarrassment, a sad and unsettling end. The title to the last chapter of almost every book about Bryan features the word "retreat" or "decline."

The minority view, gaining ground in recent biographies and clearly correct in my judgment, holds
that Bryan never transformed or retreated, and that he
viewed his last battle against evolution as an extension
of the populist thinking that had inspired his life's work.
But how can a move to ban the teaching of evolution
in public schools be deemed progressive, and how did
Bryan link his previous efforts to this new strategy?

Bryan's attitude to evolution rested upon a threefold error. First, he made the common mistake of
confusing the fact of evolution with the Darwinian explanation of its mechanism. He then misinterpreted
natural selection as a martial theory of survival by battle
and destruction of enemies. Finally, he fell into the
logical error of arguing that Darwinism implied the
moral virtuousness of such deathly struggle. The first
two errors may count as simple misunderstandings of a
theory within the magisterium of science. But the crucial third error, the source of Bryan's emotional and
political commitment, represents his confusion of scientific with moral truth—a basic violation of NOMA,
and the foundation of almost all our unnecessary strife
over evolution and ethics. Bryan wrote in the *Prince of
Peace* (1904):

The Darwinian theory represents man as reaching his present perfection by the operation of

the law of hate—the merciless law by which the
strong crowd out and kill off the weak. If this is
the law of our development then, if there is any
logic that can bind the human mind, we shall
turn backward toward the beast in proportion
as we substitute the law of love. I prefer to believe that love rather than hatred is the law of
development.

In 1906, Bryan told the sociologist E. A. Ross that
"such a conception of man's origin would weaken the
cause of democracy and strengthen class pride and the
power of wealth." He persisted in this uneasiness until
World War I, when two events galvanized him into
frenzied action. First, he learned that the martial view
of Darwinism had been invoked by most German intellectuals and military leaders as a justification for war and
future domination. Second, he feared the growth of
skepticism at home particularly as a source of moral
weakness in the face of German militarism.

Bryan united these new fears with all his previous
doubts into a campaign against evolution in the classroom. We may question the quality of his argument,
but we cannot deny that his passion on this subject
arose from his lifelong zeal for progressive causes. Consider the three principal foci of his campaign, and their
links to his populist past:

1. For peace and compassion against militarism and murder. "I learned," Bryan wrote, "that it was Darwinism that was at the basis of that damnable doctrine that might makes right that had spread over Germany."

2. For fairness and justice toward farmers and workers and against exploitation for monopoly and profit. Darwinism, Bryan argued, had convinced so many entrepreneurs about the virtue of personal gain that government now had to protect the weak and poor from an explosion of anti-Christian moral decay: "In the United States," he wrote,

> pure-food laws have become necessary to keep manufacturers from poisoning their customers; child labor laws have become necessary to keep employers from dwarfing the bodies, minds and souls of children; anti-trust laws have become necessary to keep overgrown corporations from strangling smaller competitors, and we are still in a death grapple with profiteers and gamblers in farm products.

3. For the rule of majority opinion against imposing elites. Christian belief still enjoyed widespread majority support in America, but college education was eroding a consensus that had once ensured compassion

within democracy. Bryan cited studies showing that only 15 percent of college male freshmen harbored doubts about God, but that 40 percent of graduates had become skeptics. Darwinism, and its immoral principle of domination by a selfish elite, had fueled this skepticism. Bryan railed against this insidious undermining of morality by a minority of intellectuals, and he vowed to fight fire with fire. If they worked through the classroom, he would respond in kind and ban their doctrine from the public schools. The majority of Americans did not accept human evolution, and had a democratic right to proscribe its teaching.

Let me pass on this third point. Bryan's contention strikes at the heart of academic freedom, and scientific questions cannot be decided by majority vote in any case. I merely record that Bryan embedded his curious argument in his own concept of populism. "The taxpayers," he wrote,

> have a right to say what shall be taught . . . to direct or dismiss those whom they employ as teachers and school authorities . . . The hand that writes the paycheck rules the school, and a teacher has no right to teach that which his employers object to.

But what of Bryan's first two arguments about the influence of Darwinism on militarism and domestic exploitation? We may detect the touch of the Philistine in Bryan's claim, but we must also admit that he had identified something deeply troubling—and that the fault does lie partly with violations of NOMA by scientists and their acolytes.

Bryan often stated that two books had altered the character of his opposition to evolution from laissez-faire to vigorous action: *Headquarters Nights,* by Vernon L. Kellogg (1917), and *The Science of Power*, by Benjamin Kidd (1918). I read these two books and found them every bit as riveting as Bryan had. I also came to understand his fears, even to agree in part (though not, of course, with his analysis or his remedies).

Vernon Kellogg was an entomologist and perhaps the leading teacher of evolution in America (he held a professorship at Stanford and wrote a major textbook, *Evolution and Animal Life,* with his mentor and Darwin's leading disciple in America, David Starr Jordan, ichthyologist and president of Stanford University). During the First World War, while America maintained official neutrality, Kellogg became a high official in the international, nonpartisan effort for Belgian relief, a cause officially "tolerated" by Germany. In this capacity, he was posted at the headquarters of the German Great General Staff, the only American on the premises. Night after night he listened to dinner discussions and arguments, sometimes in the presence of the Kaiser himself, among Germany's highest military officers. *Headquarters Nights* is Kellogg's account of these exchanges. He arrived in Europe as a pacifist, but left committed to the destruction of German militarism by force.

Kellogg was appalled, above all, at the justification for war and German supremacy advanced by these officers, many of whom had been university professors before the war. They not only proposed an evolutionary rationale but advocated a false and particularly crude version of natural selection, defined as inexorable, bloody battle:

> Professor von Flussen is Neo-Darwinian, as are most German biologists and natural philosophers. The creed of the *Allmacht* ["all might," or omnipotence] of a natural selection based on violent and competitive struggle is the gospel of the German intellectuals; all else is illusion and anathema.
>
> . . . This struggle not only must go on, for that is the natural law, but it should go on so that this natural law may work out in its cruel, inevitable way the salvation of the human species . . . That human group which is in the most advanced evolutionary stage . . . should win in the struggle for existence, and this struggle should occur precisely that the various types

may be tested, and the best not only preserved, but put in position to impose its kind of social organization—its *Kultur*—on the others, or, alternatively, to destroy and replace them. This is the disheartening kind of argument that I faced at Headquarters . . . Add the additional assumption that the Germans are the chosen race, and that German social and political organization is the chosen type of human community life, and you have a wall of logic and conviction that you can break your head against but can never shatter—by headwork. You long for the muscles of Samson.

Kellogg, of course, found in this argument only "horrible academic casuistry and . . . conviction that the individual is nothing, the state everything." Bryan conflated a perverse interpretation, based on a fundamental violation of NOMA, with the thing itself and affirmed his worst fears about the polluting power of evolution.

Benjamin Kidd, an English commentator highly respected in both academic and lay circles, wrote several popular books on the implications of evolution. In *The Science of Power* (1918), his posthumous work, Kidd constructs a curious argument that, in a very different way from Kellogg's, also fueled Bryan's dread. Kidd, a philosophical idealist, believed that life can only progress by rejecting material struggle and individual benefit. Like the German militarists, but to excoriate rather than to praise, Kidd identified Darwinism with domination by force. He argued, for example, that Darwinism had rekindled the most dangerous of human tendencies—our pagan soul, previously (but imperfectly) suppressed for centuries by Christianity and its doctrines of love and renunciation:

> The hold which the theories of the *Origin of Species* obtained on the popular mind in the West is one of the most remarkable incidents in the history of human thought ... Everywhere throughout civilization an almost inconceivable influence was given to the doctrine of force as the basis of legal authority ...
>
> For centuries the Western pagan had struggled with the ideals of a religion of subordination and renunciation coming to him from the past. For centuries he had been bored almost beyond endurance with ideals of the world presented to him by the Churches of Christendom ... But here was a conception of life which stirred to its depths the inheritance in

him from past epochs of time . . . This was the world which the masters of force comprehended. The pagan heart of the West sang within itself again in atavistic joy.

We may conclude that Bryan was playing, probably quite unconsciously, the classic game of "blaming the victim" in excoriating Darwin, or the theory of natural selection, or even evolution itself, as the chief source of moral decay in his time. The originator of an idea cannot be held responsible for egregious misuse of his theory (unless such misuse arises from the originator's own confusion or poor expression, which he then, in a fit of pique or haughtiness, makes no effort to correct; it just isn't Alexander Graham Bell's fault that your teenage kid's phone bill nearly bankrupted you last year). Bryan, as noted previously, failed to comprehend evolution in nearly all conceivable ways. He certainly did not understand Darwin's idea of natural selection, which is not a principle of victory by mortal combat, but a theory about reproductive success, however that goal be best achieved in local environments (by combat in some circumstances, to be sure, but by cooperation in others). But, most important, for the context of this book, Bryan never grasped NOMA's chief principle that factual truth, however constituted, cannot dictate, or even imply, moral truth. Any argument that facts or theories of biological evolution can enjoin or validate any moral behavior represents a severe misuse of Darwin's great insight, and a cardinal violation of NOMA.

But Bryan continued to characterize evolution as a principle of battle and destruction of the weak, a doctrine that undermined any decent morality and deserved banishment from the classroom. In a rhetorical flourish near the end of his "Last Evolution Argument," the final speech that he prepared with great energy, but never had an opportunity to deliver at the Scopes trial, Bryan proclaimed:

> Again force and love meet face to face, and the question "What shall I do with Jesus?" must be answered. A bloody, brutal doctrine— Evolution—demands, as the rabble did nineteen hundred years ago, that He be crucified.

I wish I could stop here with a snide comment on Bryan as Yahoo and a ringing defense for science's proper interpretation of Darwinism. But such a dismissive judgment would be unfair, because Bryan cannot be faulted on one crucial issue. Lord only knows, he understood precious little about science, and he wins no medals for logic of argument. But when he said that Darwinism had been widely portrayed as a defense of war, domination, and domestic exploitation, he was right.

We now come to the crux of this story. Such misuses of Darwinism stand in violation of NOMA, and have also perpetrated much mischief in our century. But who bears the responsibility for such misuse? If scientists had always maintained proper caution in their interpretations, and proper humility in resisting invalid extensions of their findings into the inappropriate domains of other magisteria, then we could exonerate my profession by recognizing the inevitable misuses by nonscientists as yet another manifestation of the old adage that no good deed goes unpunished.

But NOMA cuts both ways and imposes restriction and responsibility on both magisteria. The political campaigns of American creationists do represent—as usually and correctly interpreted—an improper attempt by partisans of a marginal and minority view within the magisterium of religion to impose their doctrines upon the magisterium of science. But, alas, scientists have also, indeed frequently, been guilty of the same offense in reverse, even if they don't build organized political movements with legislative clout.

Many people believe that evolution validates this or that moral behavior because scientists have told them so. When we view the behavior thereby justified as either benign or harmless, we tend to look the other way, and give the scientist a pass for his hubris. But fashions change, and today's benevolence may become tomorrow's anathema. The average American male reader in 1900 probably accepted racism, with his group on top, as a dictate of nature, and probably supported imperial expansion of American power. The claim that evolution justified the morality of both conclusions probably seemed, to him, both evident and reasonable. And if a prominent biologist advanced such a statement, then the argument became even more persuasive.

Most people today—from the subsequent perspective of Ypres, Hiroshima, lynchings, and genocides—consider such transgressions from evolutionary fact to social morality as both insidious and harmful. Bryan drew a valid lesson from his reading. Several of the German generals who traded arguments with Kellogg had been university professors of biology. Scientists cannot claim immunity from misinterpretations, particularly from socially harmful arguments advanced in violation of NOMA, if their own colleagues become frequent proposers and perpetrators.

Let me close with a specific example from a chillingly relevant source. In his "Last Evolution Argument," Bryan charged that evolutionists had misused science to present moral opinions about the social order as though they represented facts of nature.

By paralyzing the hope of reform, it discourages those who labor for the improvement of

man's condition . . . Its only program for man
is scientific breeding, a system under which
a few supposedly superior intellects, self-
appointed, would direct the mating and the
movements of the mass of mankind—an impossible system!

Who can fault Bryan here? One of the saddest
chapters in the entire history of science records the extensive misuse of data to support the supposed moral
and social consequences of biological determinism, the
claim that inequalities based on race, sex, or class cannot
be altered because they reflect the innate and inferior
genetic endowments of the disadvantaged. Enough
harm has been done by scientists who violate NOMA
by misidentifying their own social preferences as facts of
nature in their technical writings. How much more
mischief might arise when scientists who write textbooks, particularly for elementary and high-school students, promulgate these social doctrines as the objective
findings of their profession.

I own a copy of the book that John Scopes used to
teach evolution to the children of Dayton, Tennessee—
A Civic Biology, published in 1914 by George William
Hunter, professor of biology at Knox College. Many
writers have consulted this book to find the sections on

evolution that Scopes taught and Bryan quoted. But I also discovered disturbing comments in other chapters that have eluded previous commentators—an egregious claim, for example, that science holds the moral answer to questions about mental retardation, or social poverty so misinterpreted. Hunter discusses the infamous Jukes and Kallikaks, the "classic," and false, cases once offered as canonical examples of how bad heredity runs in families. Under the heading "Parasitism and Its Cost to Society—The Remedy," Hunter writes:

> Hundreds of families such as those described above exist today, spreading disease, immorality and crime to all parts of this country. The cost to society of such families is very severe. Just as certain animals or plants become parasitic on other plants and animals, these families have become parasitic on society. They not only do harm to others by corrupting, stealing or spreading disease, but they are actually protected and cared for by the state out of public money. Largely for them the poorhouse and the asylum exist. They take from society, but they give nothing in return. They are true parasites.
>
> If such people were lower animals, we would probably kill them off to prevent them

from spreading. Humanity will not allow this, but we do have the remedy of separating the sexes in asylums or other places and in various ways preventing intermarriage and the possibilities of perpetuating such a low and degenerate race.

In another passage, just two pages after the famous diagram that Bryan held aloft to demonstrate how Scopes had taught the insidious notion that humans might be classified as mammals, Hunter writes a single paragraph under the heading "the races of man"—in a textbook assigned to children of all groups in public high schools throughout America:

At the present time there exist upon the earth five races or varieties of man, each very different from the other in instincts, social customs, and, to an extent, in structure. These are the Ethiopian or negro type, originating in Africa; the Malay or brown race, from the islands of the Pacific; the American Indian; the Mongolian or yellow race, including the natives of China, Japan, and the Eskimos; and finally, the highest type of all, the Caucasians, represented by the civilized white inhabitants of Europe and America.

Bryan advocated the wrong solution, but he had correctly identified a serious problem!

Science is a discipline, and disciplines are exacting. All disciplines maintain rules of conduct and self-policing. All gain strength, respect, and acceptance by working honorably within their bounds and knowing when transgression upon other realms counts as hubris or folly. Science, as a discipline, tries to understand the factual state of nature and to explain and coordinate these data into general theories. Science teaches us many wonderful and disturbing things—facts that need weighing when we try to develop standards of conduct, and when we ponder the great questions of morals and aesthetics. But science cannot answer these questions alone and science cannot dictate social policy.

Scientists have power by virtue of the respect commanded by the discipline. We may therefore be sorely tempted to misuse that power to further a personal prejudice or social goal: Why not provide that extra oomph by extending the umbrella of science over a personal preference in ethics or politics? But we cannot, lest we lose the respect that tempted us in the first place. NOMA cuts both ways.

We live with poets and politicians, preachers and philosophers. All have distinctive ways of knowing, valid in their proper domains. No single way can hold all the answers in our wondrously complex world. Besides, highfalutin morality aside, if we continue to overextend the boundaries of science, folks like Bryan will nail us properly for their own insidious purposes.

We should give the last word to Vernon Kellogg, a great teacher who understood the principle of strength in limits, and who listened with horror to the ugliest misuses of Darwinism. Kellogg properly taught in his textbook (with David Starr Jordan) that Darwinism cannot provide moral answers:

> Some men who call themselves pessimists because they cannot read good into the operations of nature forget that they cannot read evil. In morals the law of competition no more justifies personal, official, or national selfishness or brutality than the law of gravitation justifies the shooting of a bird.

To which, let all people of goodwill; all who hold science, or religion, or both, dear; all who recognize NOMA as the logically sound, humanely sensible, and properly civil way to live in a world of honorable diversity—let them say, Amen.

4

PSYCHOLOGICAL REASONS FOR CONFLICT

Can Nature Nurture Our Hopes?

F OR TRADITIONALISTS OF THE OLD order, 1859 was
not the best of years. The principal mark and symbol must lie, inevitably and permanently (at least so long
as our culture endures), with the publication of Darwin's
Origin of Species. But Darwin's vision of a morally neutral world, not constructed for human delectation (and
not evidently cognizant either of our presence or our
preferences for comfort), received an unusual boost from
the literary sensation of the same year—the first edition
of Edward Fitzgerald's very free translation of the
Rubaiyat of Omar Khayyam, the eleventh-century Persian
mathematician and freethinker. Each of Omar's quatrains embodies a philosophical gem of resignation to a
world without intrinsic sense or desired form. (*Rubaiyat*
is the plural of *ruba'i*, a distinctive four-line form of
verse with rhymes on the first, second, and fourth lines.)

Instead of presenting the conventional quotes from Darwin, some lines from Omar may give us even more insight into the angst of mid-Victorian times, as traditional moral certainties eroded before a juggernaut of technological transformation and colonial expansion, all fueled by the progress of science. Consider this thought on the cosmic confusion of it all:

> Into this Universe, and Why not knowing,
> Nor whence, like Water willy-nilly flowing:
> And out of it, as Wind along the Waste
> I know not Whither, willy-nilly blowing.

Or this on the earth's mean estate (a shabby hotel for camel caravans!) and the meandering nature of our lives:

> Think, in this battered Caravanserai
> Whose Portals are alternate Night and Day,
> How Sultan after Sultan with his Pomp
> Abode his destined Hour, and went his way.

Or this on our inability to make nature conform to our hopes and dreams:

> Ah Love! could you and I with Fate conspire
> To grasp this sorry Scheme of Things entire,

Would we not shatter it to bits—and then
Remold it nearer to the Heart's Desire!

Why should we not, in such a world, "take the cash, and let the credit go," to cite Omar's most enduring line (usually misattributed to Adam Smith, J. M. Keynes, Donald Trump, or some other figure from a more immediate Western world).

This book rests on a basic, uncomplicated premise that sets my table of contents and order of procedure, and that requires restatement at several points in the logic of my argument: NOMA is a simple, humane, rational, and altogether conventional argument for mutual respect, based on non-overlapping subject matter, between two components of wisdom in a full human life: our drive to understand the factual character of nature (the magisterium of science), and our need to define meaning in our lives and a moral basis for our actions (the magisterium of religion).

I sketch this argument, with examples of support from leaders on both sides, in the first two chapters. The second half of the book then examines the central paradox of why such an eminently sensible solution to the nonproblem of supposed conflict between science and religion—a resolution supported by nearly all major thinkers in both magisteria—has been poorly comprehended and frequently resisted. The two major reasons, defining the last two chapters of this book, can also be simply stated and understood—even if the actual history of discussion, based on a chronicle of confusion, has been downright byzantine. I treated the first, or *historical*, reason in chapter 3: the reluctance of many religious devotees to withdraw from turf once legitimately occupied under previous views of life and nature, but now properly deeded to the newer magisterium of science (combined with the symmetrical imperialism of many scientists who stage similar invalid forays into the magisterium of moral argument).

I now devote this final chapter to the second, or *psychological*, reason—an issue whose stark simplicity should also stand forth, even in the historical morass of actual struggle: we live in a vale of tears (or at least on a field of confusion), and we therefore clutch at any proffered comfort of an encompassing sort, however dubious the logic, and however contrary the evidence.

I opened this chapter with classic doubts, from an eleventh-century Persian poet, about nature's beneficence. We may consult an equally classic Western source for the complement to this fear about nature— our anxiety about our own status and our ability to make sense of our surroundings. Consider these famous lines (heroic couplets rather than quatrains this time) from Alexander Pope's *Essay on Man* (1733–34):

Placed on this isthmus of a middle state,
A being darkly wise and rudely great . . .
He hangs between; in doubt to act or rest;
In doubt to deem himself a god, or beast . . .
Created half to rise, and half to fall;
Great lord of all things, yet a prey to all;
Sole judge of truth, in endless error hurl'd;
The glory, jest, and riddle of the world!

Such compounded anxiety about nature and human understanding must generate "rescue fantasies," to cite a catchphrase of contemporary therapy. We long to situate ourselves on a benevolent, warm, furry, encompassing planet, created to provide our material needs, and constructed for our domination and delectation. Unfortunately, this pipe dream of succor from the realm of meaning (and therefore under the magisterium of religion) imposes definite and unrealistic demands upon the factual construction of nature (under the magisterium of science). But nature, who is as she is, and who existed in earthly form for 4.5 billion years before we arrived to impose our interpretations upon her, greets us with sublime indifference and no preference for accommodating our yearnings. We are therefore left with no alternative. We must undertake the hardest of all journeys by ourselves: the search for meaning in a place both maximally impenetrable and closest to home—within our own frail being.

We should therefore, with grace and optimism, embrace NOMA's tough-minded demand: Acknowledge the personal character of these human struggles about morals and meanings, and stop looking for definite answers in nature's construction. But many people cannot bear to surrender nature as a "transitional object"—a baby's warm blanket for our adult comfort. But when we do (for we must), nature can finally emerge in her true form: not as a distorted mirror of our needs, but as our most fascinating companion. Only then can we unite the patches built by our separate magisteria into a beautiful and coherent quilt called wisdom.

The misguided search for intrinsic meaning within nature—the ultimate (and also the oldest) violation of NOMA—has taken two principal forms in Western traditions. I call the first approach the "Psalm Eight," or "all things under his feet," solution, to commemorate both the honest and accurate posing of the question: How, in the light of our cosmic smallness, can we even contemplate any favorable intrinsic meaning?

When I consider thy heavens, the work of thy fingers; the moon and the stars which thou hast ordained; What is man, that thou art mindful of him?

and also the boastful answer of our vainglorious dreams:

For thou hast made him a little lower than the angels, and hast crowned him with glory and honor,

followed by the false construction of nature, as previously quoted (see page 109):

Thou hast put all things under his feet: all sheep and oxen, yea, and the beasts of the field; the fowl of the air, and the fish of the sea.

In other words, "all things under his feet" finds meaning in nature by touting our superiority over other creatures, or advocating the more extreme position that nature exists to serve our needs. If this first solution focuses on the human side, the second strategy of "all things bright and beautiful" identifies warmth, fuzziness, and moral rectitude as the unambiguous pattern of nature. If we wish to integrate ourselves into this ennobling totality, we must, in the closing words of the parable of the Good Samaritan, go and do likewise.

Both of these "all things" solutions founder upon nature's intransigence. The solutions, recording our hopes for domination and solace, require that nature be constructed in a particular way. But nature resists our implied architecture by flaunting a set of contrary factual patterns discovered within the magisterium of science. (Under NOMA, these contrary facts do not confute "religion," or even the prospect of a religious conception of nature; they speak only against particular interpretations advanced by some religious people, and by many nonreligious folks as well.)

I will not rehearse *in extenso* the familiar arguments against "all things under his feet"—see my previous book *Full House*, or almost any contemporary volume on principles of evolution or the diversity of life. *Homo sapiens* may be the brainiest species of all, but we represent only a tiny twig, grown but yesterday on a single branch of the richly arborescent bush of life. This bush features no preferred direction of growth, while our own relatively small limb of vertebrates ranks only as one among many, not even as *primus inter pares*. *Homo sapiens* is a single species among some two hundred species of primates, on a branch of some four thousand species of mammals, on a limb of nearly forty thousand species of vertebrates, on a bough of animals dominated by more than a million described species of insects. The other boughs of life's bush have longer durations and greater prospects for continued success—while bacteria build the main trunk and have always dominated the history of life by criteria of diversity, flexibility, range of habitats and modes of life, and sheer weight of numbers.

The complementary fallacy of "all things bright and

beautiful" may be illustrated by the standard example from classical literature on natural history—a case glossed over by supporters, but squarely faced by Darwin, and therefore providing a segue to the next section on Darwin's seminal defense of NOMA against the psychological impediment.

In fairness, honorable supporters of "all things bright and beautiful" have always recognized that they cannot prove their case with furry pandas, gaudy butterflies, or the noble solicitude of Bambi's father. For the contrary argument does not deny that some creatures charm our aesthetic sense or evoke our moral approbation (because we have read their overt actions in the inappropriate light of human judgment, not because we have understood the evolutionary basis of such behaviors for the creatures themselves—often an entirely different matter). But, *prima facie,* nature also seems replete with behaviors that our moral traditions would label as ugly and cruel. And these frequent cases of ostensible opposition, not the familiar examples of apparent support, set the challenge that "all things bright and beautiful" must overcome if advocates really wish to argue that the moral meaning of life lies exposed in nature's factuality. For if we allow nature to define morality, then we must either claim that nature's ways embody traditional values of love, kindness, and cooperation—or we must admit that Kellogg's German generals were right after all, that the Golden Rule and the Ten Commandments represent unattainable fantasies, and that the moral order includes frequent murder and rapine.

The obstacles faced by "all things bright and beautiful" are steep indeed. Just consider Darwin's incisive argument that most cases of apparent support record an opposite reality when we dig deeper:

> We behold the face of nature bright with gladness, we often see superabundance of food; we do not see, or we forget, that birds which are idly singing round us mostly live on insects or seeds, and are thus constantly destroying life; or we forget how largely these songsters, or their eggs, or their nestlings, are destroyed by birds and beasts of prey.

Therefore, advocates for nature's intrinsic goodness had to find a straight and narrow path down a street with dangers on both sides. Against one flank, they needed to reassert the traditional interpretation of conventional appearances in the face of Darwin's counterargument, quoted above. But, to avoid the other flank, they had to face the even more difficult task of convincing people that cases of nature's apparent ugliness really embody moral rectitude when understood in a deeper sense.

In mounting a defense for such an improbable argument, supporters of "all things bright and beautiful" adopted the "ichneumonid wasp" (a group of several hundred species, not just a single creature) as a test case. In translation to human values, the reproductive behavior of these insects could not possibly be more disgusting, or more grisly. The mother wasp seeks another insect, usually a caterpillar, as a host for her young. She then either injects her eggs into the host's body, or paralyzes the host with her sting and then lays the eggs on top. When the eggs hatch, the larvae eat the living, often paralyzed, host from the inside—but very carefully, leaving the heart and other vital organs for last, lest the host decay and spoil the bounty. (In the spirit of false comparison, we might analogize this behavior with the old punishment of drawing and quartering for treason, a procedure devised for the same grim purpose of postponing death to extract maximal torture.) J. M. Fabre, the most famous entomological writer of Darwin's century, described the situation in his customarily graphic manner:

> One may see the cricket, bitten to the quick, vainly move its antennae and abdominal styles, open and close its empty jaws, and even move a foot, but the larva is safe and searches its vitals with impunity. What an awful nightmare for the paralyzed cricket!

Now, how can "all things bright and beautiful" be defended in the face of such horrendous realities (in the inappropriate light of human judgment, to be sure, but "all things bright and beautiful" explicitly shines such a light as its central premise)? Several resolutions have been proposed by scientists who, denying NOMA, wish to assert that nature's facts can set a foundation for human morality. Consider three examples, all from leading naturalists of Darwin's time, not from marginal figures.

1. The paralyzed hosts may suffer, and the whole system isn't very nice, but nature exists for humanity, and any device for human benefit records nature's good intentions. For example, Charles Lyell, in his great textbook on *Principles of Geology* (1830–33), argued that any natural check upon noxious insects, including the death of many as hosts to wasp larvae, could only record nature's construction for human benefit since these insects might destroy our agriculture "did not Providence put causes in operation to keep them in due bounds."

2. Some features of the system may seem to fall on the downside of moral worth, but, considered as a totality, the good guides for human conduct greatly outnumber the bad. William Kirby, Rector of Barham and Britain's leading entomologist, waxed poetic about the

love demonstrated by caring mothers in provisioning infants they would never see:

> A very large proportion of them are doomed to die before their young come into existence. But in these the passion is not extinguished . . . When you witness the solicitude with which they provide for the security and sustenance of their future young, you can scarcely deny to them love for a progeny they are never destined to behold.

Kirby also put in a good word for the marauding larvae, praising them for their forbearance in eating selectively to keep their caterpillar alive. Would we all husband our resources with such care!

> In this strange and apparently cruel operation one circumstance is truly remarkable. The larva of the Ichneumon, though every day, perhaps for months, it gnaws the inside of the caterpillar, and though at last it has devoured almost every part of it except the skin and intestines, carefully all this time it avoids injuring the vital organs, as if aware that its own existence depends on that of the insect upon which it preys! . . . What would be the impression which a similar

instance amongst the race of quadrupeds would make upon us? If, for example, an animal . . . should be found to feed upon the inside of a dog, devouring only those parts not essential to life, while it cautiously left uninjured the heart, arteries, lungs, and intestines,—should we not regard such an instance as a perfect prodigy, as an example of instinctive forbearance almost miraculous?

3. The paralyzed, but pulsating, caterpillars surely seem to suffer in thrashing agony, but we have been misled. First of all, the wriggling of the caterpillar arises as a mechanical consequence of movement by the foraging larvae inside! Second, lower animals are automata and feel no pain. St. George Mivart, an eminent critic of Darwin, argued that "many amiable and excellent people" had been misled by the apparent suffering of animals. Using a favorite racist argument of the time— that "primitive" people suffer far less than advanced and cultured folk—Mivart extrapolated further down the ladder of life into a realm of very limited pain indeed. Physical suffering, he argued,

depends greatly upon the mental condition of the sufferer. Only during consciousness does it exist, and only in the most highly organized men does it reach its acme. The author has been assured that lower races of men appear less keenly sensitive to physical suffering than do more cultivated and refined human beings. Thus only in man can there really be any intense degree of suffering, because only in him is there that intellectual recollection of past moments and that anticipation of future ones, which constitute in great part the bitterness of suffering. The momentary pang, the present pain, which beasts endure, though real enough, is yet, doubtless, not to be compared as to its intensity with the suffering which is produced in man through his high prerogative of self-consciousness.

No one has ever matched Mark Twain for roasting scientific arrogance, particularly when extended into areas (like morality) where science has no business. In a satire titled "Little Bessie Would Assist Providence," Twain chronicles a family conversation. The daughter insists that a benevolent God would not have given her little friend "Billy Norris the typhus" or visited other unjust disasters upon decent people. Her mother responds that there must be a good reason for it all. Bessie's last rejoinder, which summarily ends the essay, invokes the ultimate and classical case of the ichneumons:

Mr. Hollister says the wasps catch spiders and cram them down into their nests in the ground—alive, mama!—and there they live and suffer days and days and days, and the hungry little wasps chewing their legs and gnawing into their bellies all the time, to make them good and religious and praise God for His infinite mercies. I think Mr. Hollister is just lovely, and ever so kind; for when I asked him if he would treat a spider like that he said he hoped to be damned if he would; and then he—Dear mama, have you fainted!

In 1860, after reading the *Origin of Species*, Asa Gray wrote to Charles Darwin, explaining (as discussed on pages 197–203) that he could accept natural selection as God's mode of action, but that he still felt compelled to find a moral purpose behind all evolutionary results. Darwin responded, in his wonderfully honest way, that he could not, as a scientist, resolve issues about moral purposes and ultimate meanings—but that he simply could not imagine how nature's factual particulars could possibly be squared with traditional values. Interestingly, he cited two examples of behaviors that can only be judged as intensely disturbing if we analyze them (wrongly, Darwin insisted) in terms of human moral values—a common (and troubling) observation of many pet owners, and the less familiar but ultimately grisly "standard" of the ichneumons:

> I own that I cannot see as plainly as others do, and as I should wish to do, evidence of design and beneficence on all sides of us. There seems to me too much misery in the world. I cannot persuade myself that a beneficent and omnipotent God would have designedly created the Ichneumonidae with the express intention of their feeding within the living bodies of Caterpillars, or that a cat should play with mice.

In their different styles, Darwin and Twain provided the proper response, and rang the death knell over "all things bright and beautiful"—indeed, over any false argument that seeks the basis of moral truth (or any other concept under the magisterium of religion, including the nature and attributes of God) in the factual construction of the natural world. NOMA demands separation between nature's factuality and humankind's morality—dare I say that never the Twain shall meet?

The ichneumonid story is nothing but horrendous when rendered in our ethical terms. But framing such a factual issue "in our terms" cannot be defended in a natural world neither made for us nor ruled by us—and quite incapable, in any case, of providing any moral instruction for human propriety. The devouring of living and paralyzed caterpillars is an evolutionary strategy that works for ichneumons, and that natural selection has programmed into their behavioral repertoire. Caterpillars are not suffering to teach us something; they have simply been outmaneuvered, for now, in the evolutionary game. Perhaps they will evolve a set of adequate defenses sometime in the future, thus sealing the fate of ichneumons. And perhaps, indeed probably, they will not.

Nature's Cold Bath and Darwin's Defense of NOMA

D ARWIN HAS BEEN READ AS something of a moral dolt, or at least as a slacker on the subject, for his frequent disclaimers about drawing lessons for the meaning of human life from his revolutionary reorganization of biological knowledge. Shouldn't such a radical reinterpretation of nature offer us some guidance for the biggest questions of the ages: Why are we here, and what does it all mean? How could anyone look so deeply into the heart of biological causality and the history of life, and then offer us such a piddling dribble—*bupkes,* as my grandmother would have said—on the meaning of life and the ultimate order of things:

> I feel most deeply that the whole subject is too profound for the human intellect. A dog might as well speculate on the mind of Newton.

Was Darwin just a coward? a desiccated intellect? a small-minded man? the very stereotype of a scientist who can describe a tree and ignore the forest, or analyze the notes and not hear the symphony?

I view Darwin in an entirely opposite manner. He maintained, throughout his life, a basic human fascination for the great questions of morals and meanings, and he recognized the transcendent importance of such inquiry. But he knew both the strengths and the limitations of his chosen profession, and he understood that the power of science could only be advanced and consolidated on the fertile ground of its own magisterium. In short, Darwin rooted his views about science and morality in the principle of NOMA.

Darwin did not use evolution to promote atheism, or to maintain that no concept of God could ever be squared with the structure of nature. Rather, he argued that nature's factuality, as read within the magisterium of science, could not resolve, or even specify, the existence or character of God, the ultimate meaning of life, the proper foundations of morality, or any other question within the different magisterium of religion. If many Western thinkers had once invoked a blinkered and indefensible concept of divinity to declare the impossibility of evolution, Darwin would not make the same arrogant mistake in the opposite direction, and claim that the fact of evolution implies the nonexistence of God.

I would go further and argue that we have often, and seriously, misconstrued Darwin's basic view about proper relationships between nature and the meaning of human life. Darwin's position, rooted in NOMA, is courageous, tough-minded, and ultimately liberating. But we have often misread his vision as defeatist, pessimistic, and enslaving. I propose that we call Darwin's view the "cold bath" theory of nature.

The basic argument includes three propositions linked in a definite order of implication:

1. The basic statement of NOMA. The facts of nature are what they are, and cannot, in principle, resolve religious questions about God, meaning, and morality.

2. Two alternatives for nature. Unconstrained by our religious hopes and needs, nature remains free to assume any appearance when read in the invalid light of human moral or aesthetic judgment. Consider two extreme possibilities and the different temptations they inspire. Perhaps, by sheer good fortune, nature generally does follow our preferences for warmth and fuzziness. Perhaps most organisms are cute or beautiful in our sight, and perhaps peaceful cooperation does usually prevail over violent competition. Perhaps Isaiah's holy mountain, where the wolf dwells with the lamb and the

leopard lies down with the kid, does generally record nature's factual estate, not only our idyllic dreams.

At the other extreme, perhaps nature rarely matches our hopes. Perhaps we can counterpose an ugly tapeworm to every beautiful peacock; an ichneumon larva foraging within a living caterpillar to every dolphin who ever raised an incapacitated relative into breathable air; an evolutionary triumph of adaptation by loss of complexity in an amorphous parasite to every evolutionary triumph of adaptation by increased braininess in a human ancestor.

By the logic of NOMA, the potential validity of either extreme doesn't make a particle of difference. We still cannot draw moral messages or religious conclusions from any factual construction of nature—either from extreme warmth and fuzziness, or from maximal distastefulness. But we all recognize the primary foible of frail humanity—our propensity for embracing hope and shunning logic, our tendency to believe what we desire rather than what we observe. In the light of this weakness, we will be sorely tempted to make a serious mistake if, under the first extreme, the facts of nature do tend, generally and fortuitously, to match our desires. We will then be beguiled into violating NOMA and rushing headlong into the error of conflating these facts with values and meanings. Wouldn't we be better off if nature, for equally fortuitous reasons, happens to refute our hopes and desires most of the time?

3. Better an invigorating cold bath than a suffocating warm embrace. Nature is amoral—not immoral, but rather constructed without reference to this strictly human concept. Nature, to speak metaphorically, existed for eons before we arrived, didn't know we were coming, and doesn't give a damn about us. Thus, it would be passing strange if the first extreme held and nature generally reflected our moral and aesthetic preferences. The odds against such a coincidence—the accidental match of an independent system to an entirely different and equally complex institution originating so much later—must be astronomically high.

In fact, and by all honest reckoning, such a match does not exist. Nature conforms to neither extreme of human definition. Nature betrays no statistical preference for being either warm and fuzzy, or ugly and disgusting. Nature just is—in all her complexity and diversity, in all her sublime indifference to our desires. Therefore we cannot use nature for our moral instruction, or for answering any question within the magisterium of religion. We certainly cannot follow the old, intellectually squishy tradition of searching for moral certainties within nature's supposedly warm and fuzzy ways. We cannot even accept the arch and opposite argument of T. H. Huxley, who held, in his most famous essay (*Evolution and Ethics,* 1893) that, because the rules of evolution violate all standards of human ethical conduct, the moral lesson of nature must be sought in learning her patterns and then behaving in a precisely contrary manner!

> The practice of that which is ethically best— what we call goodness or virtue—involves a course of conduct which, in all respects, is opposed to that which leads to success in the cosmic struggle for existence. In place of ruthless self-assertion it demands self-restraint; in place of thrusting aside, or treading down, all competitors, it requires that the individual shall not merely respect, but shall help his fellows . . . It repudiates the gladiatorial theory of existence . . . Laws and moral precepts are directed to the end of curbing the cosmic process.

Instead, Darwin argues, we must simply admit that nature offers no moral instruction at all. We must, in other words, take the ultimate cold bath of immersing ourselves in nature and recognizing that, for this particular quest, we have come to the wrong place. Such a "cold bath" may shock us at first. But as we experience the invigoration of such a bracing surround, we should come to view the immersion as neither grim nor depressing, but exhilarating and liberating. If we then stop searching for moral truth in material reality, we may finally appreciate nature's fascination and her extensive powers to resolve different, but equally important, questions within her own realm. And when we reject the siren song of false sources, we become free to seek solutions to questions of morals and meanings in the proper place—within ourselves.

I noted in the first chapter (page 35) that I regarded Darwin's letter to Asa Gray as the finest statement ever written on the proper relation of factual nature to human morality or, more broadly, of science to religion. I now return to the extended logic of Darwin's argument in formulating the "cold bath" theory of nature as the liberating principle of NOMA. Darwin, we recall, begins by disclaiming on the meaning of evolution for theological questions—except in refuting the old delusion that an intrinsically beneficent nature records God's existence and attributes:

> With respect to the theological view of the question. This is always painful to me. I am bewildered. I had no intention to write atheistically. But I own that I cannot see as plainly as others do, and as I should wish to do, evidence of design and beneficence on all sides of us.

How, then, shall we interpret nature's facts, particularly those (like foraging ichneumons, and cats "playing" with battered mice) that we view with horror in our inappropriate moral terms:

> I am inclined to look at everything as resulting from designed laws, with the details, whether good or bad, left to the working out of what we may call chance.

Two points of this subtle argument deserve special notice. First, Darwin may accept overall design as a personal preference, or even as a guide to his own existence and comfort, but he knows that such issues cannot be adjudicated within the magisterium of science—as expressed in his later suspicion that such questions are "too profound for the human intellect." Second, Darwin makes a clear distinction between such scientifically unknowable ultimates and particular events and patterns (nature's factuality) that can be described and explained within the magisterium of science. Then— following the chief precept of NOMA—Darwin denies that we may hope to locate, in these factual events, either the hand of God or a moral lesson for the conduct of our lives. I particularly value the insight and precision of Darwin's words: "with the details, whether good or bad, left to the working out of what we may call chance."

Darwin does not mean "chance" in the vernacular

senses of "random," "without meaning," or "incapable of explanation." By stating the proviso "what we may call chance," he implicates a view of life for which he had no word, but which historians now call "contingency." That is, nature's facts (the "details") exist for immediate, definite, and potentially knowable reasons subject to scientific explanation. But these facts are not integrated into any controlling fabric of a planned and deterministic universe, with intended meaning in the fall of each petal and every raindrop.

The universe, for all we know, may have an ultimate purpose and meaning ("I am inclined to look at everything as resulting from designed laws"), and these ultimates may be set by a rational and transcendent power legitimately called God, but the resolvable subject matter of science falls into another realm below the purview of such philosophical (and probably unknowable) generalities. Moreover, these smaller and knowable facts unfold in a world composed of so many complex parts that prediction of the future, not to mention inferences about ultimate meanings of the totality, cannot be achieved with certainty. We may use the laws of nature, and our knowledge of specific conditions, to explain and understand particular events, and even (as the highest goal of science) to construct general theories about factual patterns in nature. We can know "what" and "how," even "why" in the special sense of explaining particular facts by invariant laws of nature and properties of materials. But science has no access to questions of ultimate "why" expressed as overarching purpose or eternal value.

To show that I am not presenting an odd and personal exegesis of Darwin's statement about "details, whether good or bad, left to the working out of what we may call chance," we may trace his own explication in a beautifully crafted set of examples, moving from undeniable commonplaces to challenging implications that we would rather not accept—all designed to convince his conventionally devout colleague, Asa Gray.

Darwin moves slowly and cautiously, but ever so systematically. If a man, caught on top of a hill during a thunderstorm, dies by lightning, the event surely has a scientific explanation—based on general laws (of meteorology and electricity) and particular conditions (the man's location at a given moment). But no one would claim either that the man's death could have been predicted with precision at the time of his birth (or even an hour before his demise), or, especially, that the tragedy occurred for a reason rooted in good morals and the ultimate meaning of things. The poor fellow was just in the wrong place at the wrong time—while nature, morally blind as always, followed her usual rules. Darwin writes: "The lightning kills a man, whether a good one or bad one, owing to the excessively complex action of natural laws."

If such a tragic natural death has no moral meaning, what about a tragic natural birth? Darwin next argues that a mentally handicapped child may owe his condition to rules of genetics and embryology, as applied to the particular circumstances of his being. His condition can therefore be explained scientifically. But only a moral pervert could believe that the child's handicap was meant to be because it happened, or that God follows an agenda of overall decency by purposely peppering our lives with such specific misfortune. Darwin writes: "A child (who may turn out an idiot) is born by the action of even more complex laws." ("Idiot," in Darwin's day, was a technical term for a definite level of mental deficiency, not a label of opprobrium.)

Darwin has now reached the crux of his argument: the births and deaths of individuals may be explained naturally, but such scientific reasons do not imply either necessary occurrence in a deterministic universe, or moral meaning under God's omnipotence. At this point a believer in the old order, preferring God's moral presence in factual events to NOMA's insistence on the separation of magisteria, might say, "Fine, God doesn't busy himself with the fates of individuals; he grants this space to the ancient doctrine of 'free will.' But God surely controls larger patterns and generalities for moral ends. He may allow the birth of an individual to fall outside his purview, but he will not so neglect the birth of an entire species, especially not the origin of *Homo sapiens*, the apple of his eye, the incarnation of his image, and the ultimate goal of all that came before."

Darwin, who has been setting Gray up for this denouement all along, now moves in for the kill. If a single baby is only an individual in a population of human beings, why should a single species rank as any more than an individual among all earthly species in the fullness of geological time? And why should *Homo sapiens* be viewed as a goal and a generality, when *Pharkidonotus percarinatus* (a favorite fossil snail of mine—I am not making this name up), which lived for a much longer time with markedly larger populations, ranks only as a particular accident of history? What, beyond our dangerous and unjustified arrogance, could even permit us to contemplate such a preferred status for one species among the hundreds of millions that have graced the history of our planet? Therefore human existence must also be judged as a "detail . . . left to the working out of what we may call chance." And we have already agreed, for the man killed by lightning and the child born with severe handicaps, that such details cannot embody moral messages or reveal ultimate meanings. Darwin writes: "I can see no reason why a man, or other animal, may not have been aboriginally produced by other laws."

Darwin wrote this letter to Gray on May 22, 1860. Gray's response elicited a further statement from Darwin in July—an even more forceful argument that nature's facts, even the ones we like best (especially the origin of our own species), cannot reveal God's purposes or life's ultimate meaning:

> One more word on "designed laws" and "undesigned results." I see a bird which I want for food, take my gun and kill it, I do this designedly. An innocent and good man stands under a tree and is killed by a flash of lightning. Do you believe (and I really should like to hear) that God designedly killed this man? . . . If you believe so, do you believe that when a swallow snaps up a gnat that God designed that that particular swallow should snap up that particular gnat at that particular instant? I believe that the man and the gnat are in the same predicament. If the death of neither man nor gnat are designed, I see no good reason to believe that their *first* birth or production should be necessarily designed.

To people who find such a cold bath depressing, who feel that the quality of human life must be degraded and cheapened in a universe without intrinsic meaning recorded in our terms, and who fear that our inability to glean moral truth from the facts of nature can only lead to a destructive ethical relativism (or even to a denial of the existence or importance of morality at all), I can only urge the wisdom behind the opposite reading, as championed by Darwin and embodied in NOMA.

What can be more deluding, or even dangerous, than false comfort that blinds our vision and inspires passivity? If moral truth lies "out there" in nature, then we need not struggle with our own confusions, or with the varying views of fellow humans in our diverse world. We can adopt the much more passive approach of observing nature (or just accepting what "experts" tell us about factual reality) and then aping her ways. But if NOMA holds, and nature remains neutral (while bursting with relevant information to spice our moral debates), then we cannot avoid the much harder, but ultimately liberating, task of looking into the heart of our distinctive selves.

I do not deny the comfort of older views that fractured NOMA and defined the universe in terms of our hopes and supposed powers. "All things under his feet" may fuel the body, as "all things bright and beautiful" fuels the soul. But this food may be poison under a sugarcoating. The combined wisdom of all classes in all cultures—from the pomp of past power reduced to legs of stone in the desert of Shelley's "Ozymandias," to the common fate of palookas ("the bigger they are, the harder they fall")—proclaims the virtue of tough-minded modesty, and the location of real power in realms of appropriate and effective action.

To anyone who feels cosmically discouraged at the prospect of life as a detail in a vast universe not evidently designed for our presence, I offer two counter-arguments and an item of solace. Consider, first, the much greater fascination and intellectual challenge of such a mysterious but knowable universe, compared with a "friendlier" and more familiar cosmos that only mirrors our hopes and needs. Then contemplate, secondly, the happier prospect of fulfilling the Socratic dictum to "know thyself" by actively trying to fathom a distinctively human nature within, rather than passively imbibing a generalized external nature without, as we struggle to define the purposes of our lives.

Finally, and for solace, I present a wonderful sonnet by Robert Frost, so tightly keyed to Darwin's argument in his letters to Gray (another quintessential New Englander) that I must locate Frost's inspiration in his intimate knowledge of Darwin's writing (as expressed in several other poems as well). Frost, on a morning walk, encounters an odd conjunction of three white objects with different geometries. This peculiar but fitting combination, he argues, must record some form of intent; it cannot be accidental. But if intent be truly manifest, than what can we make of our universe—for the scene is evil by any standard of human morality. We must take heart in Darwin's proper solution: We are really observing one of those "details" that, "whether good or bad," belongs to the domain "of what we may call chance." Design does not govern here:

I found a dimpled spider, fat and white,
On a white heal-all, holding up a moth
Like a white piece of rigid satin cloth—
Assorted characters of death and blight
Mixed ready to begin the morning right,
Like the ingredients of a witches' broth—
A snow-drop spider, a flower like a froth,
And dead wings carried like a paper kite.

What had that flower to do with being white,
The wayside blue and innocent heal-all?
What brought the kindred spider to that height,
Then steered the white moth thither in the night?
What but design of darkness to appall?—
If design govern in a thing so small.

Homo sapiens also ranks as a "thing so small" in a vast universe, a wildly improbable evolutionary event, and not the nub of universal purpose. Make of such a conclusion what you will. Some people find the prospect depressing. I have always regarded such a view of life as exhilarating—a source of both freedom and consequent moral responsibility. We are the offspring of history, and must establish our own paths in this most diverse and interesting of conceivable universes—one indifferent to our suffering, and therefore offering us maximal freedom to thrive, or to fail, in our own chosen way.

The Two False Paths of Irenics

I'M ALWAYS OPEN TO A new word. Lord knows we invent enough of them within my domain of science. A few years ago, I came across a theological term that tickled my fancy, both for its touch of the arcane, and its mellifluous ring—*irenics* (from the Greek word for "peace"), defined in opposition to polemics as a branch of Christian theology that "presents points of agreement among Christians with a view to the ultimate unity of Christianity" *(Oxford English Dictionary)*. By extension (and the word has crept out of theological circles and into general English usage), irenic people and proposals "tend to promote peace, especially in relation to theological and ecclesiastical differences."

Now, I'm an irenic fellow at heart—and I trust that most of us so regard ourselves, whatever personal quirks and foibles stand in the way of realization. This book promotes an irenic solution under a large umbrella extending far beyond the purely Christian realm of official definitions cited above. I join nearly all people of goodwill in wishing to see two old and cherished institutions, our two rocks of ages—science and religion—coexisting in peace while each works to make a distinctive patch for the integrated coat of many colors that will celebrate the distinctions of our lives, yet cloak human nakedness in a seamless covering called wisdom.

Irenics sure beats the polemics of ill-conceived battle between science and religion—a thoroughly false model (chapter 2) that too often continues to envelop us for illogical reasons of history (chapter 3) and psychology (chapter 4). I do get discouraged when some of my colleagues tout their private atheism (their right, of course, and in many ways my own suspicion as well) as a panacea for human progress against an absurd caricature of "religion," erected as a straw man for rhetorical purposes. Religion just can't be equated with Genesis literalism, the miracle of the liquefying blood of Saint Januarius (which at least provides an excuse for the wonderful and annual San Gennaro Festival on the streets of New York), or the Bible codes of kabbalah and modern media hype. If these colleagues wish to fight superstition, irrationalism, philistinism, ignorance, dogma, and a host of other insults to the human intellect (often politically converted into dangerous tools of murder and oppression as well), then God bless them—but don't call this enemy "religion."

Similarly (of course), I pronounce my anathema upon those dogmatists and "true believers" who, usurping the good name of religion for their partisan doctrines, try to suppress the uncomfortable truths of science, or to impose their peculiar brand of moral fiber upon people with legitimately different tastes. Careers are short, and while I won't deny some good moments of comedy, and even of prideful achievement, I'd sure rather be studying the evolution and paleontology of West Indian land snails than fighting creationists. 'Nuff said.

If we embrace the alternative premise that irenicism should prevail between science and religion, then what form should our peaceful interaction take? In making this volume's closing argument for NOMA as the most honorable, and also the most fruitful, form of irenics, I wish to revisit an important principle of intellectual life, previously discussed in highfalutin terms as Aristotle's golden mean between extremes (see page 49), but here embodied in the "Goldilocks principle" of "just right" between too much and too little, too soft and too hard, or too hot and too cold. NOMA represents the bed of proper firmness, and the right amount of oatmeal at the right temperature. NOMA honors the sharp differences in logic between scientific and religious arguments.

NOMA seeks no false fusion, but urges two distinct sides to stay on their own turf, develop their best solutions to designated parts of life's totality, and, above all, to keep talking to each other in mutual respect, and with an optimistic forecast about the value of reciprocal enlightenment. In other words, citing Churchill's aphorism, to "jaw-jaw rather than war-war."

This Goldilocks solution provides the right firmness of extensive contact with respect for inherent differences, the right amount of dialogue for devotees of disparate subjects, and the right temperature of discourse for inputs that do not blend. Call NOMA irenics with a punch. The dialogue will be sharp and incisive at times; participants will get riled up, as a blessed consequence of our unextinguishable human nature; but respect for legitimate differences, and a recognition that full answers require distinctive contributions from each side, should maintain a field of interest, honor, and productive struggle.

On the important theme of enemies within versus enemies without, the anti-irenic conflict-mongers who violate NOMA by trying to expand their side into the other's magisterium pose a greater threat under conventional notions about overt opposition. But they also display the general virtue of "enemies without": we know where they stand, and we know how to fight back. However, among those who preach irenicism, two prominent approaches would undermine NOMA from within by seeking peace between science and religion under strategies that paralyze the principles of NOMA. I view these two alternate irenicisms as the extremes within a common domicile (the house of peace, in this case) that Goldilocks rejected for a middle way.

The first alternative—too hot, too soft, and too much—continues to amaze me by persistence, even growth, in the face of massive internal contradictions that should have driven such a misguided notion to extinction ages ago. This *syncretic* school continues to embrace the oldest fallacy of all as a central premise: the claim that science and religion should fuse to one big, happy family, or rather one big pod of peas, where the facts of science reinforce and validate the precepts of religion, and where God shows his hand (and mind) in the workings of nature. (The word *syncretic* includes both admirable and unfavorable meanings. In choosing such a name for this overwrought style of irenicism, I had only the negative definitions in mind (of *Webster's Third New International Dictionary*): "flagrant compromise in religion or philosophy; eclecticism that is illogical or leads to inconsistency; uncritical acceptance of conflicting or divergent beliefs or principles.")

Much as modern syncretism riles me, I can at least take comfort in a wry feature of the contemporary version—at least from the parochial perspective of a professional scientist. Older and classical forms of syncretism always gave the nod to God—that is, religion set the outlines that everyone had to accept, and science then had to conform. Irenics in this older mode required that the principles and findings of science yield religious results known in advance to be true. Indeed, such conformity represented the primary test of science's power and validity. For example, Thomas Burnet (see pages 16–25) did not doubt that the biblical narrative recorded the earth's actual history; his scientific job, by his lights, required validation of this known history in terms of causation by invariant natural laws rather than miracles.

But the spectacular growth and success of science has turned the tables for modern versions of syncretism. Now the conclusions of science must be accepted a priori, and religious interpretations must be finessed and adjusted to match unimpeachable results from the magisterium of natural knowledge! The Big Bang happened, and we must now find God at this tumultuous origin.

I'm sorry. I know that I shouldn't be so dismissive, especially (and ironically) in a section about irenicism. But I find the arguments of syncretism so flawed, so illogical, so based in hope alone, and so freighted by past procedures and certainties, that I have difficulty keeping a straight face or a peaceful pen.

I also feel particularly sensitive about this issue because, as I wrote this book in the summer of 1998, a deluge of media hype enveloped the syncretist position, as though some startlingly new and persuasive argument had been formulated, or some equally exciting and transforming discovery had been made. In fact, absolutely nothing of intellectual novelty had been added, as the same bad arguments surfaced into a glare of publicity because the J. M. Templeton Foundation, established by its fabulously wealthy eponym to advance the syncretist program under the guise of more general and catholic (small c) discussion about science and religion, garnered a splash of media attention by spending 1.4 million bucks to hold a conference in Berkeley on "science and the spiritual quest."

In a genuine example of true creation *ex nihilo*—that is, the invention of an issue by fiat of media reports, rather than by force of argument or content of material—at least three major sources preached the syncretist gospel in their headlines and vapidly uncritical reports: "Faith and Reason, Together Again" (*The Wall Street Journal,* June 12); "Science and Religion: Bridging the Great Divide" (*The New York Times,* June 30); and a cover story in *Newsweek* (July 20) simply titled "Science Finds God." Scientists could only be mystified by this last claim, but at least we can now be certain about one of God's attributes: he sells newspapers and magazines.

The *Times* article admitted the intellectual torpor of the proceedings: "A kind of Sunday school politeness pervaded the meeting, with none of the impassioned confrontations expected from such an emotionally charged subject . . . The audience politely applauded after each presentation. But there was little sense of intellectual excitement." But from whence could such excitement arise in principle? If NOMA holds (and I devote this book to advocating the validity of this proposition), then facts and explanations developed under the magisterium of science cannot validate (or deny) the precepts of religion. Indeed, if we look at the so-called arguments for syncretism, as described in these reports, they all devolve into a series of fuzzy statements awash in metaphor and illogic. Consider just three examples, not chosen as egregiously silly, but representative of the standard fare.

1. Woolly metaphor misportrayed as decisive content. *Newsweek* reports the following fusion of Christ and quanta:

Take the difficult Christian concept of Jesus as both fully divine and fully human. It turns out that this duality has a parallel in quantum physics. In the early years of this century, physicists discovered that entities thought of as particles, like electrons, can also act as waves . . . The orthodox interpretation of this strange situation is that light is, simultaneously, wave and particle . . . So, too, with Jesus, suggests physicist F. Russell Stannard of England's Open University. Jesus is not to be seen as really God in human guise, or as really human but acting divine, says Stannard: "He was fully both."

Now what am I to make of such a claim? That the status of Jesus as both God and man (a central trinitarian concept) must be factually true because electrons, and other basic components, can be construed as either waves or particles? I don't see what such a comparison could indicate except that the human mind can embrace contradiction (an interesting point, to be sure, but not a statement about the factual character of God), and that people can construct the wildest metaphors.

2. Clutching at straws based on superficial similarity. *The Wall Street Journal* presents the following two remarkable examples of the syncretist notion that science can validate spiritual claims. We learn, first of all, that Darwin himself was a closet syncretist:

Surprisingly, among the first to reunite science and religion was Darwin. The participants argued that he destroyed the notion of God as an absent clock-winder, restoring the ever-present deity of the Psalms. As Arthur Peacocke put it, Darwin permitted "a recovered emphasis on ancient insights," showing that "God is creating all the time."

Again, what am I supposed to conclude from such fuzziness? Has the factuality of an old-fashioned creating God been proven because Darwin used developmental language to describe the genealogical history of life? I thought that the God of many Christians confined this kind of creative activity to the early days of life's history. Or is Mr. Peacocke's God just retooling himself in the spiffy language of modern science?

We next learn that Genesis finds confirmation in the latest developments of cosmology:

> The Big Bang, now believed to have taken place
> 15 billion years ago, accords neatly enough with
> Genesis.

Now what, pray tell, is "neatly enough"? Some folks insist that Genesis occurred less than ten thousand years ago. Moreover, the Big Bang cannot be touted as a description of God's initial creation of the universe *ex nihilo*. The Big Bang does not set the ultimate beginning of all material things—a subject outside the magisterium of science. The Big Bang is a proposition about the origin of our *known* universe. This scientific theory cannot, in principle, specify what, if anything, happened before (if such a notion even has any meaning)—because any previous history gets erased when and if the stuff of the universe collapses to such an effective point of origin.

3. Plain, old-fashioned illogic. The *pièce de résistance* of modern syncretism, at least in almost all public accounting I have ever read, lies in the so-called anthropic principle—a notion with as many definitions as supporters, and which, in my view, is either utterly trivial in its "weak versions" (the designation of supporters, not my deprecation), or completely illogical in "strong versions." *The Wall Street Journal* explains the anthropic principle as "the biggest hint" of God's presence in the findings of science:

> What this means is that complex, carbon-based life—namely us—can exist only in a universe in which the physical constants have been tuned just so. Take the ratio of gravity to electromagnetism. If gravity were a tiny bit stronger, we'd be pulled apart; if electromagnetism were a tiny bit stronger, we'd fall in on ourselves like failed soufflés.

Yes, but so what? The weak version only tells us that life fits well with nature's laws, and couldn't exist if the laws were even the tiniest bit different. Interesting, but I see no religious implications—and, in fairness, neither do most syncretists (thus their own designation as "weak"). The "strong" version provides my favorite example of illogic in high places. Since human life couldn't exist if the laws of nature were even a tad different, then the laws must be as they are because a creating God desired our presence.

This argument reduces to pure nonsense based on the unstated premise—which then destroys the "strong anthropic principle" by turning it into a classic example of circular reasoning—that humans arose for good and necessary reasons (and that whatever allowed us to get here must therefore exist to fulfill our destiny). Without this premise (which I regard as silly, arrogant, and utterly unsupported), the strong anthropic principle collapses upon the equal plausibility of this opposite interpretation: "If the laws of nature were just a tad different, we wouldn't be here. Right. Some other configuration of matter and energy would then exist, and the universe would present just as interesting a construction, with all parts conforming to reigning laws of this different nature. Except that we wouldn't be around to make silly arguments about this alternate universe. So we wouldn't be here. So what." (I'm glad we are here, by the way—but I don't see how any argument for God's existence can emerge from my pleasure.)

Readers may have laughed at the old and absurd arguments I cited for the divine benevolence of ichneumons feeding upon live and paralyzed caterpillars (see page 183). You may have wondered why I chose to devote so much space to such a straw-man violation of NOMA from a bad old past, now superseded. But will future generations view these current syncretist arguments against NOMA, and for the inference of God from facts of nature, as any wiser?

The second irenicist alternative to NOMA—too cold, too hard, and too little—requires only a paragraph or two of commentary because no intellectual argument, but only current (and lamentable) social custom, fuels the strategy. The syncretists may be silly, but at least they talk and try. The opposite irenicism of "no offense, please, we're politically correct" adopts the fully avoidant tactic of never generating conflict by never talking to each other, or speaking in such muted and meaningless euphemisms that no content or definition can ever emerge. Sure, we can avoid the language of racial conflict if we vow never to talk about race. But what then can change, and what can ever be resolved?

And, yes, we could bring science and religion into some form of coexistence under political correctness if all scientists promised never to say anything about religion, and all religious professionals swore that the troublesome S-word would never pass their lips. Contemporary American culture has actually adopted this unholy contract for many issues that should be generating healthy debate, and surely cannot ever be brought to a fair conclusion if we don't talk to each other. Intellectuals can only regard such voluntary suppression of discussion as a guarantee that tough but resolvable issues will continue to fester and haunt us, and as a sin—I don't know how else to say this—against the human mind and heart. If we have so little confidence in our unique mental abilities, and in our intrinsic goodwill, then what indeed is man (and woman) that anyone should be mindful of us?

NOMA does cherish the separate status of science and religion—regarding each as a distinctive institution, a rock for all our ages, offering vital contributions to human understanding. But NOMA rejects the two paths to irenicism on either side of its own tough-minded and insistent search for fruitful dialogue—the false and illogical union of syncretism, and the perverse proposal of "political correctness" that peace may best be secured by the "three monkeys" solution of covering eyes, ears, and mouth.

The non-overlapping magisteria of science and religion must greet each other with respect and interest on the most distinctively human field of talk. To close with a rationale from each magisterium, scientists generally argue that language represents the most special and transforming feature of human distinctiveness—and only a dolt would fail to lead with his best weapon. As for religion, this book began with the story of Doubting Thomas from the end of John's gospel. So let me take a page from *Finnegans Wake*, and become recursive by ending this book with the beginning of the same document. I do know, of course, that the phrase bears another meaning in its original context, but John also acknowledged the same precious uniqueness—the key to resolving our conflicts, and the positive force behind NOMA—in starting his gospel with a true guide to salvation: In the beginning was the Word.

Index

An *n* following a page number indicates an item found in a note at the bottom of the page.

ABOUT THE AUTHOR

The author of more than fifteen books, STEPHEN JAY GOULD is also author of the longest-running contemporary series of scientific essays, which appears monthly in *Natural History*. He is the Alexander Agassiz Professor of Zoology and professor of geology at Harvard; is curator for invertebrate paleontology at the university's Museum of Comparative Zoology; and serves as the Vincent Astor Visiting Professor of Biology at New York University. He lives in Boston, Massachusetts, and New York City.

Now available from
THE LIBRARY OF
CONTEMPORARY THOUGHT

VINCENT BUGLIOSI
NO ISLAND OF SANITY
Paula Jones v. Bill Clinton
The Supreme Court on Trial

JOHN FEINSTEIN
THE FIRST COMING
Tiger Woods: Master or Martyr?

PETE HAMILL
NEWS IS A VERB
Journalism at the End of the Twentieth Century

CARL HIAASEN
TEAM RODENT
How Disney Devours the World

SEYMOUR M. HERSH
AGAINST ALL ENEMIES
Gulf War Syndrome: The War Between America's
Ailing Veterans and Their Government

EDWIN SCHLOSSBERG
INTERACTIVE EXCELLENCE
Defining and Developing New Standards
for the Twenty-first Century

ANNA QUINDLEN
HOW READING CHANGED MY LIFE

WILLIAM STERLING AND
STEPHEN WAITE
BOOMERNOMICS
The Future of Your Money
in the Upcoming Generational Warfare

JIMMY CARTER
THE VIRTUES OF AGING

SUSAN ISAACS
BRAVE DAMES AND WIMPETTES
What Women Are Really Doing on Page and Screen

HARRY SHEARER
IT'S THE STUPIDITY, STUPID
Why (Some) People Hate Clinton
and Why the Rest of Us Have to Watch

Coming from
**THE LIBRARY OF
CONTEMPORARY THOUGHT**
*America's most original writers
give you a piece of their minds*

Robert Hughes
Jonathan Kellerman
Joe Klein
Walter Mosley
Donna Tartt
Don Imus
Nora Ephron

**Look for these titles coming soon from
The Library of Contemporary Thought**

ROBERT HUGHES
A JERK ON ONE END
Reflections of a Mediocre Fisherman

JONATHAN KELLERMAN
SAVAGE SPAWN
Reflections on Violent Children